Remote Sensing the Mekong

The Mekong Basin in Southeast Asia is one of the largest international river basins in the world. Its abundant natural resources are shared by six riparian countries and provide the basis for the livelihoods of more than 75 million people. However, ongoing socioeconomic growth and related anthropogenic interventions impact the region's ecosystems, and there is an urgent need for the monitoring of the basin's land surface dynamics. Remote sensing has evolved as a key tool for this task, allowing for up-to-date analyses and regular monitoring of environmental dynamics beyond physical or political boundaries and at various temporal and spatial scales. This book serves as a forum for remote-sensing scientists with an interest in the Mekong River Basin to present their recent basin-related works as well as applied case studies of the region. A broad range of sensors from high to medium resolution, and from multispectral to radar systems, are applied, covering topics such as land cover/land use classification and comparison, time series analyses of climate variables, vegetation structure and vegetation productivity, and studies on flood mapping or water turbidity monitoring. This book was originally published as a special issue of the *International Journal of Remote Sensing*.

Claudia Kuenzer is the Head of the Land Surface Department at the German Remote Sensing Data Center (DFD) of the German Aerospace Center (DLR). She has led several research projects on the Mekong region and published numerous journal articles, book chapters, and a book on the Mekong Delta.

Patrick Leinenkugel is the Head of the team Coasts and River Basins at the German Remote Sensing Data Center (DFD) of the German Aerospace Center (DLR). He has contributed to multiple research initiatives in the Mekong Basin and has published several remote sensing studies for the region.

Stefan Dech is the Director of DLR's German Remote Sensing Data Center (DFD), and University Professor and holder of the endowed chair in remote sensing in the Department of Geography at Julius-Maximilians-Universität Würzburg, Germany. He has co-authored numerous Mekong-related papers.

Remote Sensing the Mekong

Edited by
Claudia Kuenzer, Patrick Leinenkugel, and Stefan Dech

Routledge
Taylor & Francis Group

LONDON AND NEW YORK

First published 2017 by Routledge

2 Park Square, Milton Park, Abingdon, Oxfordshire OX14 4RN
52 Vanderbilt Avenue, New York, NY 10017

*Routledge is an imprint of the Taylor & Francis Group, an informa
business*

First issued in paperback 2018

British Library Cataloguing in Publication Data
A catalogue record for this book is available from the British Library

ISBN 13: 978-0-415-30638-6 (hbk)
ISBN 13: 978-0-367-13906-3 (pbk)

Typeset in TimesNewRomanPS
by diacriTech, Chennai

Publisher's Note
The publisher accepts responsibility for any inconsistencies that may have arisen
during the conversion of this book from journal articles to book chapters, namely
the possible inclusion of journal terminology.

Disclaimer
Every effort has been made to contact copyright holders for their permission to
reprint material in this book. The publishers would be grateful to hear from any
copyright holder who is not here acknowledged and will undertake to rectify any
errors or omissions in future editions of this book.

Contents

CONTENTS

Citation Information

The chapters in this book were originally published in the *International Journal of Remote Sensing*, volume 35, issue 8 (April 2014). When citing this material, please use the original page numbering for each article, as follows:

Chapter 1
Preface: Remote sensing the Mekong
Claudia Kuenzer
International Journal of Remote Sensing, volume 35, issue 8 (April 2014) pp. 2747–2751

Chapter 2
Comparing global land-cover products – implications for geoscience applications: an investigation for the trans-boundary Mekong Basin
Claudia Kuenzer, Patrick Leinenkugel, Matthias Vollmuth, and Stefan Dech
International Journal of Remote Sensing, volume 35, issue 8 (April 2014) pp. 2752–2779

Chapter 3
Land-surface temperature dynamics in the Upper Mekong Basin derived from MODIS time series
C.M. Frey and C. Kuenzer
International Journal of Remote Sensing, volume 35, issue 8 (April 2014) pp. 2780–2798

Chapter 4
Sensitivity analysis for predicting continuous fields of tree-cover and fractional land-cover distributions in cloud-prone areas
Patrick Leinenkugel, Michel L. Wolters, Claudia Kuenzer, Natascha Oppelt, and Stefan Dech
International Journal of Remote Sensing, volume 35, issue 8 (April 2014) pp. 2799–2821

Chapter 5
Recent climate variability and its impact on precipitation, temperature, and vegetation dynamics in the Lancang River headwater area of China
Chong Huang, Yafei Li, Gaohuan Liu, Hailong Zhang, and Qingsheng Liu
International Journal of Remote Sensing, volume 35, issue 8 (April 2014) pp. 2822–2834

Chapter 6

Drought impact on vegetation productivity in the Lower Mekong Basin
Binghua Zhang, Li Zhang, Huadong Guo, Patrick Leinenkugel, Yu Zhou, Li Li, and Qian Shen
International Journal of Remote Sensing, volume 35, issue 8 (April 2014) pp. 2835–2856

Chapter 7

Suitability of SAR imagery for automatic flood mapping in the Lower Mekong Basin
Felix Greifeneder, Wolfgang Wagner, Daniel Sabel, and Vahid Naeimi
International Journal of Remote Sensing, volume 35, issue 8 (April 2014) pp. 2857–2874

Chapter 8

Ecosystem assessment in the Tonle Sap Lake region of Cambodia using RADARSAT-2 Wide Fine-mode SAR data
Lu Zhang, Huadong Guo, Xinwu Li, and Liyan Wang
International Journal of Remote Sensing, volume 35, issue 8 (April 2014) pp. 2875–2892

Chapter 9

The Ha Tien Plain – wetland monitoring using remote-sensing techniques
Tim Funkenberg, Tran Thai Binh, Florian Moder, and Stefan Dech
International Journal of Remote Sensing, volume 35, issue 8 (April 2014) pp. 2893–2909

Chapter 10

Operational multi-sensor monitoring of turbidity for the entire Mekong Delta
Thomas Heege, Viacheslav Kiselev, Magnus Wettle, and Nguyen Nghia Hung
International Journal of Remote Sensing, volume 35, issue 8 (April 2014) pp. 2910–2926

For any permission-related enquiries please visit:
http://www.tandfonline.com/page/help/permissions

Notes on Contributors

Tran Thai Binh is the Director of the GIS and Remote Sensing Research Center at the Ho Chi Minh City Institute of Resources Geography (GIRS), Vietnam Academy of Science and Technology, Vietnam.

Stefan Dech is the Director of DLR's German Remote Sensing Data Center (DFD), and University Professor and holder of the endowed chair in remote sensing in the Department of Geography at Julius-Maximilians-Universität Würzburg, Germany. He has co-authored numerous Mekong-related papers.

C.M. Frey is based at the German Remote Sensing Data Center (DFD), German Aerospace Center (DLR), Oberpfaffenhofen, Germany.

Tim Funkenberg is based at the Institute for Geography, Julius-Maximilians-Universität Würzburg, Germany.

Felix Greifeneder is a Researcher at the Institute for Applied Remote Sensing, European Academy of Bozen/Bolzano (EURAC), Bolzano, Italy.

Huadong Guo is the Director-General of the Chinese Academy of Sciences' (CAS) Institute of Remote Sensing and Digital Earth (RADI), an Academician of CAS, a Fellow of the Academy of Sciences for the Developing World (TWAS), and a Fellow of the International Eurasian Academy of Sciences (IEAS).

Thomas Heege, founder of EOMAP GmbH, has more than 20 years of experience in aquatic remote sensing and technical consultancy in manifold projects worldwide.

Chong Huang is based at the State Key Laboratory of Resources and Environmental Information System, Institute of Geographic Sciences and Natural Resources Research, Chinese Academy of Sciences, China.

Nguyen Nghia Hung is based at the Southern Institute for Water Resources Research (SIWRR), Ho Chi Minh City, Vietnam.

Viacheslav Kiselev is based at EOMAP GmbH & Co.KG, 82205 Gilching.

Claudia Kuenzer is the Head of the Land Surface Department at the German Remote Sensing Data Center (DFD) of the German Aerospace Center (DLR). She has led several research projects on the Mekong region and published numerous journal articles, book chapters, and a book on the Mekong Delta.

NOTES ON CONTRIBUTORS

Patrick Leinenkugel is the Head of the team Coasts and River Basins at the German Remote Sensing Data Center (DFD) of the German Aerospace Center (DLR). He has contributed to multiple research initiatives in the Mekong Basin and has published several remote sensing studies for the region.

Li Li is a member of staff at the College of Information Science and Engineering, Shandong Agricultural University, China.

Xinwu Li is based at the Key Laboratory of Digital Earth Science, Institute of Remote Sensing and Digital Earth, Chinese Academy of Sciences, China.

Yafei Li is based at the College of Air Traffic Management, Civil Aviation University of China, China.

Gaohuan Liu is based at the State Key Laboratory of Resources and Environmental Information System, Institute of Geographic Sciences and Natural Resources Research, Chinese Academy of Sciences, China.

Qingsheng Liu is based at the State Key Laboratory of Resources and Environmental Information System, Institute of Geographic Sciences and Natural Resources Research, Chinese Academy of Sciences, China.

Florian Moder is National Advisor at the Deutsche Gesellschaft für Internationale Zusammenarbeit (GIZ) GmbH, Germany.

Vahid Naeimi is a University Assistant at the Department for Geodesy and Geoinformation, Research Group for Remote Sensing, TU Wien, Vienna, Austria.

Natascha Oppelt is a Professor at the Institute for Geography, Christian-Albrechts-Universität Kiel, Germany.

Daniel Sabel is a Project Assistant at the Department for Geodesy and Geoinformation, Research Group for Remote Sensing, TU Wien, Vienna, Austria.

Qian Shen is based at the Laboratory of Digital Earth Science, Institute of Remote Sensing and Digital Earth, Chinese Academy of Sciences, China.

Matthias Vollmuth is a Senior Research Fellow at the School of Earth and Environmental Sciences, University of Manchester, UK.

Wolfgang Wagner is the Head of Department at the Department for Geodesy and Geoinformation, Research Group for Remote Sensing, TU Wien, Vienna, Austria.

Liyan Wang is based at the Key Laboratory of Digital Earth Science, Institute of Remote Sensing and Digital Earth, Chinese Academy of Sciences, China.

Magnus Wettle is Managing Director at EOMAP Australia Pty Ltd.

Michel L. Wolters is a Research Assistant at the German Remote Sensing Data Center (DFD), German Aerospace Center (DLR).

Binghua Zhang is based at the Key Laboratory of Digital Earth Science, Institute of Remote Sensing and Digital Earth, Chinese Academy of Sciences, China.

Hailong Zhang is based at the Key Laboratory of Remote Sensing Science, Institute of Remote Sensing and Digital Earth, Chinese Academy of Sciences, China.

NOTES ON CONTRIBUTORS

Li Zhang is based at the Key Laboratory of Digital Earth Science, Institute of Remote Sensing and Digital Earth, Chinese Academy of Sciences, China.

Lu Zhang is based at the Key Laboratory of Digital Earth Science, Institute of Remote Sensing and Digital Earth, Chinese Academy of Sciences, China.

Yu Zhou is based at the Key Laboratory of Digital Earth Science, Institute of Remote Sensing and Digital Earth, Chinese Academy of Sciences, China.

PREFACE
Remote sensing the Mekong

With a total length of over 4350 km, the Mekong River is the longest river in Southeast Asia. The river originates at the Tibetan plateau at an altitude of 5000 m and flows through six countries of the Southeast Asian mainland. The Mekong forms the political border between Laos, Myanmar, and Thailand, and further downstream marks a large part of the border between Thailand and Laos. In the lower, southern reaches, the river flows through the lowlands of Cambodia, branches off to the east in South Vietnam, and, via the Mekong Delta, empties into the South China Sea. The catchment area – the Mekong Basin – covers approximately 795,000 km^2 and supports the livelihoods of more than 72 million people. With more than 1300 fish species, the Mekong River is one of the richest fishing grounds in the world and constitutes the 'lifeline' for the inhabitants of the region. For all riparian countries, the Mekong is of outstanding national importance for agriculture (freshwater resource for irrigation), the economy (generation of energy via hydropower), food security (protein supply from fish catch and aquaculture), and trade (route of transport). Furthermore, the Mekong River is one of the elements defining people's cultural belonging.

The present political relationship between the riparian nations along the Mekong River is strongly influenced by the complex and conflict-laden historical background. Rapid economic rise and expanding economic entanglements via international trade, transnational transport, and the Mekong power grid are also accompanied by increasing regional conflicts of interest on geopolitical influence; this particularly applies to the distribution of the river's valuable water resources. Intensive land use and increasing demand for agricultural commodities have led to changes in the ecosystem and hydrological balance of the Mekong Basin. Furthermore, existing and projected large-scale dam projects in the upper reaches intensify the already precarious situation of imbalance of available water resources between neighbouring countries. Energy demand is expected to further multiply within the next decades and the river's still significant hydroelectric potential is likely to be further tapped in the future. This will have a large influence on the natural condition of the Mekong River. It can be expected that in the long term, changes in fish stocks as a result of dam projects will lead to significant losses of this resource and endanger the protein supply of the region. Consequences are expected in future land-use dynamics, with more land likely to be transformed into rangeland or aquaculture production. Already at present, land-use change rates of the region are alarming. The loss of primary and secondary growth forest, especially in Laos, Myanmar, and Cambodia, proceeds rapidly. Diverse habitats are giving way to monocultures of rubber, cashew, sugar cane, and eucalyptus, amongst others. Alongside these human-driven impacts of socio-economic transformation, urbanization, and increasing basin cooperation are influences of climate variation and natural hazards that play a role in the dynamics of the region. Overall, sustainable strategies for the development and utilization of land and water resources for the entire catchment area have a key relevance for the trans-boundary countries.

This special issue of *International Journal of Remote Sensing* serves as a forum for remote-sensing scientists with an interest in the Mekong River Basin to present their recent Basin-related works as well as applied case studies of the region. Although the Mekong Basin is of a large extent, covers numerous climate zones and eco-regions, and offers an extremely interesting playground for remote-sensing researchers from all imaginable application fields as well as from all imaginable methodological regimes, not many Earth observation-based studies on the Basin have been published in the past. Besides a small number of regional-scale remote-sensing studies undertaken for insular or continental Southeast Asia as a whole, only a few authors have published remote-sensing papers on the Mekong Delta region or case studies in other parts of the Basin, such as in northern Laos or in Yunnan, China. Overall, Basin-related studies are rare. One reason might be the fact that regional studies focusing on Southeast Asia generally do not cover the territory of China, thereby excluding the area of the Upper Mekong Basin (UMB). Another reason may be the almost persistent cloud cover over the Mekong Basin for large parts of the year. In addition, complex alpine terrain in the north of the basin and limited accessibility in other parts have constrained access for collecting ground reference data.

As an introduction to this issue, Kuenzer et al. (2014) present the physiography of the Mekong Basin, and summarize the land-cover characteristics of the region as represented in the commonly available global land-cover (GLC) products. The GLC products of the University of Maryland, UMD 1992/1993, the GLC 2000 product, the GlobCover products of 2004–2006 and 2009, as well as the Moderate Resolution Imaging Spectroradiometer (MODIS)-derived land cover products of 2001 and 2009 were compared. Even after harmonization of the legends via the Land-Cover Classification System (LCCS), cross-tabulation among the products reveals large differences. None of the current GLC products appear suitable for representing the variety and distribution of land-cover classes in the river's basin. This is of great relevance for the modelling community (climate, hydrology, biomass modellers, etc.), as well as for Mekong-related environmental studies, which should consider globally available data with caution.

Frey and Kuenzer (2014) then enter the thermal infrared domain. They analysed a 13 year time series of MODIS-derived land-surface temperature (LST) data, extracting typical annual LST patterns for the UMB. As might be expected, the southern parts of the basin exhibit higher average temperatures than the northern parts. Due to a complex topography in the north, the gradients between maximum and minimum LST, as well the gradients between daytime and night-time LST, are more accentuated in the northern parts of the watershed. The Tibetan Plateau – the origin region of the Mekong – features its own very specific regime with respect to temperature statistics, which are much more accentuated than in the margin areas of the plateau or in the Mekong highlands. Thus, even the exclusive analyses of thermal infrared time series allow for the differentiation of characteristic eco-regions. At the same time, the article critically elucidates the fact that rainy season MODIS LST observations are rare due to extensive cloud cover, and in some areas only one observation during an entire month during the rainy season is available. Thus, climate statistics for complex regions such as the Mekong have to be handled with caution. In particular, the choice of gap-filling algorithms has a large impact on the quality of a statistical product in a cloud-dominated region. Here, a good understanding of time-series handling and preprocessing will enable a potential user to understand the quality of a statistical data set.

The large proportion of cloud cover in the region also played a significant role in the sensitivity analyses for the prediction of subpixel tree cover and land cover presented by Leinenkugel et al. (2014). The authors investigated the effects of different explanatory and response variables on the accuracy and temporal robustness of predicted products, focusing on the Lower Mekong Basin (LMB), using 11 years of MODIS data as a data source. The study reveals that the prediction accuracy, which is mainly dependent on residual cloud cover in the data sets, declines markedly at the start of the rainy season. With respect to long-term monitoring purposes over several years, differences in, for example, tree-cover estimates can best be extracted when fluctuations induced by noise and external effects are minimized. In this respect the authors demonstrate that for areas with temporally concentrated cloud cover, such as those found in the Mekong Basin, intra-annual datasets do not always contribute to improved model performance but instead may negatively affect the inter-annual robustness of the predicted variables (Leinenkugel et al. 2014). The article therefore also underlines the significant impacts of data preprocessing, data handling, and model or algorithm training on the final results, which should be representative for a complex, large area.

Zhang, Zhang et al. (2014) and Huang et al. (2014) analyse medium-resolution time-series data with the goal of explaining climate-related phenomena within the basin. Zhang, Zhang et al. focus on drought events that occurred in the LMB between 2000 and 2011 and study the influence of these occurrences on vegetation productivity within the MODIS-derived MOD17 product. Especially, the strong drought of the year 2005 is well represented in the net primary productivity (NPP) data of the same year. The impacts of this drought, as well as milder droughts (albeit to a lesser extent), are especially obvious for forests, woodlands, and shrublands, whereas agricultural areas do not seem much affected. The authors postulate that in this region, drought effects are counterbalanced by irrigation and crop management techniques. On a similar note is the study by Huang et al., who investigated recent climate variability in precipitation, temperature, and vegetation dynamics datasets within the UMB. The authors identify trends in 13 years of MODIS-derived LST data, precipitation data represented by the TRMM rainfall product (3B43), and the MODIS-derived NPP product, MOD17. According to their findings, the UMB can be classified into regions by the trend in NPP over time: strongly increasing, slightly increasing, slightly decreasing, and strongly decreasing.

Whereas the above-mentioned research results focus on optical data sources, Greifeneder et al. (2014) elaborate on the suitability of SAR imagery for automatic flood mapping in the LMB. Flooding is common in most regions of the LMB – either as a reoccurring annual phenomenon, which people expect and have learned to live with, or as natural hazard events, when flood extent or duration exceeds the usual magnitude. SAR is the undisputed best source for flood mapping in cloud-prone areas. However, effects of object–sensor geometry (unfavourable incidence angles), topographic variability, weather (wind and associated increased water surface roughness), and the presence of vegetation (water–vegetation mixtures and water below canopies) can all hamper water detection. Therefore, Greifeneder et al. present a measure for the usability of SAR imagery for flood mapping in the LMB, and also present a method for the identification of permanent waterbodies. The suitability assessment, based on ENVISAT ASAR data of 150 m resolution, allows for the elimination of 90% of the scenes that are not promising for flood mapping. Furthermore, in an environment with a rainy and a dry season – such as the Mekong Basin – the authors conclude that at least an annual SAR time series is needed to derive permanent waterbodies. The more years that are included in the analysis, the higher will be the capacity for a reliable mapping of permanent waterbodies. This also means that – for the Mekong Basin – existing and upcoming permanent waterbody products, such as the permanent waterbodies product derived from the Shuttle Radar Topography Mission (SRTM) or the future TanDem-X-derived permanent waterbodies product, do not represent the true extent of permanent waterbodies for regions with seasonally fluctuating water surfaces.

One such extremely fluctuating waterbody within the Mekong Basin is the Tonle Sap Lake in Cambodia, which is Southeast Asia's largest freshwater lake and one of the largest lakes on Earth. Zhang, Guo et al. (2014) analysed the ecosystem of this Cambodian region, employing RADARSAT-2 Wide Fine Mode SAR data with dual polarization. The data, with a spatial resolution finer than 10 m, allows for the differentiation of open water surfaces, as well as flooded forest, flooded shrubland, flooded grassland, and vegetated areas that are not inundated. For such a large and complex wetland environment, SAR data is an excellent source of information – even if additional optical data might be available. Tonle Sap Lake also features unique livelihoods: the lake is home to several floating villages. In these villages, homes are built on rafts, which float freely on the lake's water surface. The study of Zhang, Guo et al. demonstrates that SAR data at the aforementioned resolution is perfectly suited for object detection, extracting these village clusters and individual houses. This SAR-based application article on a wetland is accompanied by a further application study presented by Funkenberg et al. (2014). The authors used a C5-based decision-tree classification and change-detection approach, employing multispectral Landsat data of two time periods (1991 and 2009), to depict land-cover and land-use changes in the Ha Tien Plain Wetland in the northwest of the Mekong Delta. As in many other areas of the Basin, natural wetlands and grasslands have been lost due to conversion into agriculture- and aquaculture-dominated areas.

The last article of this special issue on *Remote Sensing the Mekong* takes us from the land surface to open water. Heege et al. (2014) advance the field of water-constituent monitoring of water surfaces by presenting an operational approach for the multi-sensor monitoring of turbidity and organic matter. The study area is the entire Mekong Delta, covering the southernmost part of the Mekong Basin, where the river finally meets the ocean. The authors employ a fully automated processing chain for product generation, which can handle MODIS, Landsat, Medium Resolution Imaging Spectrometer (MERIS), Satellite Pour l'Observation de la Terre (SPOT), RapidEye, Advanced Spaceborne Thermal Emission and Reflection Radiometer (ASTER), and QuickBird data. The relevance of available ground-truth data is underlined in this article: spectral measurements of the reflective spectra of the Mekong's lower reaches allowed Heege et al. to calibrate their model for water-constituent derivation. MODIS, MERIS, and Landsat were found to be the most radiometrically stable sensors, and therefore MODIS and Landsat especially are suitable for future monitoring purposes. The algorithms used to derive water constituent products, such as turbidity and chlorphyll information, are currently employed for the entire Mekong Basin, which is of enormous significance in the hot spots of hydropower development, where it is expected that sediment loads of the river will decrease substantially after a hydropower dam is closed and in operation.

Considering the variety of contributions to this special issue, *Remote Sensing the Mekong*, its title can also stand as a synonym for the challenges that each remote-sensing scientist has to face when dealing with this region. The Mekong Basin is very large, and it exhibits a complex topography, climate, and ecosystem variety. It needs seven MODIS tiles, or over 50 Landsat frames, to cover the complete north–south-trending basin. This means that manual data processing – unless for local case studies – is near impossible. Excellent programming skills are mandatory, and for the assessment of basin dynamics, experience with time-series analyses is inevitable. Nevertheless, further Mekong-Basin-related studies are urgently required and should be encouraged. Remote-sensing-based technologies that can provide regionally consistent information across topographical gradients and across political boundaries will be crucial within this, one of the world's largest international river basins. National governments, river basin organizations, non-governmental organizations, local research institutes and universities, as well as a large community of individual scientists from diverse disciplines – all of them having the Mekong close to their heart – will gratefully make use of the results, freely available information products, and novel insights to support the region's attempts to find a balance between the ongoing economic development and the urgently needed protection of the Basin's natural resources.

The guest editor and all contributors to this special issue are grateful to the Editors-in-Chief of *International Journal of Remote Sensing* for supporting us in the realization of this issue.

Claudia Kuenzer
German Aerospace Center, DLR
Earth Observation Center, EOC
German Remote Sensing Data Center, DFD
Oberpfaffenhofen, Germany

References

Frey, C. M., and C. Kuenzer. 2014. "Land-Surface Temperature Dynamics in the Upper Mekong Basin Derived from MODIS Time Series." *International Journal of Remote Sensing* 35: 2778–2796.

Funkenberg, T., T. T. Binh, F. Moder, and S. Dech. 2014. "The Ha Tien Plain – Wetland Monitoring Using Remote-Sensing Techniques." *International Journal of Remote Sensing* 35: 2891–2907.

Greifeneder, F., W. Wagner, D. Sabel, and V. Naeimi. 2014. "Suitability of SAR Imagery for Automatic Flood Mapping in the Lower Mekong Basin." *International Journal of Remote Sensing* 35: 2855–2872.

Heege, T., V. Kiselev, M. Wettle, and N. N. Hung. 2014. "Operational Multi-Sensor Monitoring of Turbidity for the Entire Mekong Delta." *International Journal of Remote Sensing* 35: 2908–2924.

Huang, C., Y. Li, G. Liu, H. Zhang, and Q. Liu. 2014. "Recent Climate Variability and its Impact on Precipitation, Temperature and Vegetation Dynamics in the Lancang River Headwater Area of China." *International Journal of Remote Sensing* 35: 2820–2832.

Kuenzer, C., P. Leinenkugel, M. Vollmuth, and S. Dech. 2014. "Comparing Global Land-Cover Products – Implications for Geoscience Applications: An Investigation for the Trans-Boundary Mekong Basin." *International Journal of Remote Sensing* 35: 2750–2777.

Leinenkugel, P., M. L. Wolters, C. Kuenzer, N. Oppelt, and S. Dech. 2014. "Sensitivity Analysis for Predicting Continuous Fields of Tree-Cover and Fractional Land-Cover Distributions in Cloud-Prone Areas." *International Journal of Remote Sensing* 35: 2797–2819.

Zhang, L., H. Guo, X. Li, and L. Wang. 2014. "Ecosystem Assessment in the Tonle Sap Lake Region of Cambodia Using RADARSAT-2 Wide Fine-Mode SAR Data." *International Journal of Remote Sensing* 35: 2873–2890.

Zhang, B., L. Zhang, H. Guo, P. Leinenkugel, Y. Zhou, L. Li, and Q. Shen. 2014. "Drought Impact on Vegetation Productivity in the Lower Mekong Basin." *International Journal of Remote Sensing* 35: 2833–2854.

Comparing global land-cover products – implications for geoscience applications: an investigation for the trans-boundary Mekong Basin

Claudia Kuenzer[a], Patrick Leinenkugel[a], Matthias Vollmuth[b], and Stefan Dech[a]

[a]German Earth Observation Center, EOC, of the German Aerospace Center, Wessling, Germany; [b]Institute of Geography and Geology, University of Würzburg, Würzburg, Germany

In this article we present the results of a comparison of six globally available land-cover products for the Mekong Basin – an area that spans 795,000 km^2 and comprises parts of six riparian countries: China, Myanmar, Thailand, Laos, Cambodia, and Vietnam. The basin covers most climatic zones: from high-altitude, snow-covered mountainous regions in the north, to subtropical and tropical rainforest areas and agricultural land further south. The geopolitically important region not only is home to over 72,000,000 inhabitants, but also is a centre of attention of several environmental modelling experts, trying to assess future hydrologic dynamics, climate variability, as well probable land-use developments in the area.

We compare land-cover products of the University of Maryland, UMD 1992–1993, the GLC 2000 product, the GlobCover products of 2004–2006 and 2009, as well as the MODIS-derived land-cover products of 2001 and 2009. For harmonization of individual legends, the Land Cover Classification System, LCCS, has been employed. However, even after harmonization, cross-tabulation among the products reveals extreme differences, where the impact of differing classification algorithms weighs higher than the impact of temporal coincidence of products. Especially, differences within mixed-vegetation classes are large, strongly impacting the overall assessment of forested land, other vegetated land, and even cultivated land in the Mekong Basin. The findings presented here are of high relevance for the modelling community as well as Mekong-related environmental studies, which should consider global remote-sensing-derived products with caution and solid background knowledge.

1. Introduction: global land-cover mapping and products

Globally available land-cover products find their way into many geoscientific applications – be it climate-related modelling and the estimation of greenhouse gas release (Matthews 1983; Wilson and Henderson-Sellers 1985), the monitoring and analyses of drought effects (Tsegaye et al. 2010; Wang et al. 2011; Biazin and Sterk 2013), hydrologic modelling (Baker and Miller 2013; Yan et al. 2013), the derivation of net primary productivity and biomass (Eisfelder, Kuenzer, and Dech 2011), or even vulnerability and risk analyses assessing land erosion potential (Shi et al. 2013), land degradation potential (Karnieli et al. 2008; Gao and Liu 2010), agricultural potential (Baker and Griffis 2009; Schilling et al. 2010), and others. According to a survey on the relevance of remote-sensing-derived data products for the nine Societal Benefit Areas (SBAs) of the Global Earth Observation System of Systems (GEOSS), land-cover products are of the highest relevance for Earth observation user groups in the nine fields of agriculture, biodiversity, climate, disasters, ecosystems, energy, health, water, and weather (GEO

2005). The Intergovernmental Panel on Climate Change (IPCC), research groups of the United Nations (UN), or of universities and research centres, as well as large players such as the Food and Agricultural Organization (FAO), the World Health Organization (WHO), or conservation bodies such as the World Wildlife Fund (WWF) – to name only a few – are all engaged in the analyses of Earth observation data and the generation of products derived therefrom. An accentuated interest in the topic of land cover exists, and scientists, stakeholders, and decision-makers likewise underline the strong demand for high-quality products to retrieve information on the current condition of land cover and land use to reveal past changes and possible future trends impacting the state of our planet.

Land cover influences many additional land-related parameters. Land temperature strongly depends on the albedo of surface cover, as well as on the surface type's moisture content (Kuenzer and Dech 2013). Under similar illumination conditions a vegetated surface will exhibit different temperature and different diurnal temperature behaviour than a bare surface. Land cover impacts infiltration capacity and surface runoff amounts and speed (Dunjó, Pardini, and Gispert 2004; Molina et al. 2007; Peng and Wang 2012), defines patterns of evaporation (House-Peters and Chang 2011; Liu et al. 2013), defines gas uptake and release (Schneider, Grosse, and Wagner 2009; Hutyra et al. 2011; Rittenhouse and Rissman 2012; Sohl et al. 2012), influences the susceptibility to deflation and erosion (Petersen et al. 1987; Karnieli et al. 2008; Gao and Liu 2010), may define habitat boundaries for flora and fauna (Haines-Young 2009; Rojas et al. 2013), and also impacts available ecosystem services within an area (Koschke et al. 2012; Vo et al. 2012; Brown 2013). It is thus of utmost importance that the modelling community, as well as scientists integrating available land-cover information into their analyses, is supplied with products of highest possible accuracy. At the same time, modellers and researchers adopting and integrating existing land-cover products must be sensitive to the quality of their input data sets. Unfortunately, often 'data are selected without explicitly considering the suitability of the data for the specific application … and the effects of the uncertainty of the data on the results of the assessment', despite the fact that 'uncertainties due to data selection and handling can be in the same order of magnitude as uncertainties related to the representation of the process under investigation' (Verburg, Neumann, and Nol 2011, 974).

Many global and regional land-cover products are available at different spatial resolutions, ranging from 300 m to 1 km. However, the problem of comparability remains (Verburg, Neumann, and Nol 2011). This subject has been treated by selected scientists, focusing either on a general comparison between global land-cover data sets or on a comparison between global and specific regional products. Comparison is usually carried out by techniques of harmonization of different products, to achieve a common base so that agreements and disagreements of classification approaches become visible.

Latifovic et al. (2004), e.g. prepared the North and Central America land-cover product for the GLC 2000 map (GLC2000–NCA) derived from SPOT 4 VEGETATION data. For validation, the results were compared to other global land-cover products (UMD, IGBP DISCover, MODIS Land Cover Type) on an areal and pixel basis. Waser and Schwarz (2006) compared large-area land-cover products (IGBP-DISCover/UMD 1 km/ MODIS MOD12 v2) with national forest inventories and CORINE Land Cover for the region of the European Alps. In addition, Neumann et al. (2007) analysed differences between GLC2000 and CORINE 2000 land-cover data, uncovering vast disagreements. Gessner et al. (2012) compared four global land-cover maps (UMD, GLC2000, GlobCover 2004–06, and MODIS Land Cover Type 2001) after harmonization to the UN Land Cover Classification System (LCCS) for the region of West Africa to assess the influence of different land-cover data on subsequent environmental modelling. For the case of Mexico, Colditz et al. (2012) generated a specific land-cover map, derived from MODIS 250 m data, and later compared it to available global land-cover products such as MODIS Land Cover Type and GlobCover 2004–06.

Further regional studies were undertaken by Wu et al. (2008), who analysed global land-cover data sets (UMD, IGP-DISCover, MODIS Land Cover, and GLC2000) with regard to cropland area estimation in China, and Ran, Li, and Lu (2010) who compared the same global land-cover products for China, taking into account all land-cover classes. Similar to Wu et al. (2008), Fritz, See, and Rembold (2010) undertook a comparison of land-cover products (GLC2000, MODIS Land Cover, SAGE, and AFRICOVER) for Africa, focusing on the agricultural areas. Moreover, Tchuenté, Roujean, and de Jong (2011) compared land-cover products for Africa (ECOCLIMAP, GLC2000, MODIS Land Cover, and GlobCover 2004–06), but in contrast to Fritz, See, and Rembold (2010), included all land-cover types. Pérez-Hoyos, García-Haro, and San-Miguel-Ayanz (2012a) applied fuzzy comparison with the help of LCCS to compare CORINE, GLC2000, MODIS Land Cover, and GlobCover 2004–06 data for Europe. Additionally they set up a methodology to create a hybrid land-cover product for Europe, using synergies of large-scale and global land-cover products (CORINE; GLC2000, MODIS, and GlobCover 2004–06) (Pérez-Hoyos, García-Haro, and San-Miguel-Ayanz 2012b). Harmonization to LCCS was carried out to assess a common base on which to merge information of different products. Dong et al. (2012) compared a PALSAR-based forest map for South East Asia with the forest statistics of the FAO Forest Resources Assessments (FRA) 2010, GlobCover 2009, and the MODIS Land Cover 2009.

More general analyses were contributed by Hansen and Reed (2000), who compared IGBP DISCover and UMD land-cover products on a global scale while also investigating the different classification concepts. In 2006, McCallum et al. (2006) compared globally available land-cover data sets (IGBP DISCover, UMD, GLC2000, and MODIS Land Cover) to point out disagreements in the spatial distribution of land-cover classes, and a decreasing overall agreement when moving from the global to the regional scale. Jung et al. (2006) tried to take advantage of the different existing land-cover products (GLCC; GLC2000, and MODIS Land Cover Type) by creating a so-called SYNMAP, which relies on a best-estimate agreement of different maps, whereas Herold et al. (2008) introduced the approach of analysing different global land-cover maps (IGBP DISCover, UMD, MODIS Land Cover 1 km, and GLC2000) via standardized Land Cover Classification System (LCCS) harmonization. This approach is also applied in this present study.

All the cited studies share similar conclusions: the spatial resolution of global products is often not adequate for the purposes the data are to be used for. There is a strong demand for higher resolution land-cover products on a global scale, a request that the currently ongoing project for land-cover mapping based on 30 m Landsat data (Townshend et al. 2012; Gong et al. 2013; Sexton et al. 2013) addresses. Moreover, mixed classes are one main cause for low accuracies and disagreements between different products (Latifovic et al. 2004). Best agreements can be achieved for very clearly defined classes and typically in homogenous areas (Herold et al. 2006), whereas heterogeneous landscapes and transition zones constitute a major challenge to medium-resolution land-cover data (Jung et al. 2006; Verburg, Neumann, and Nol 2011). However, low accuracies also result from the differences in classification algorithms, which are the basis of the global products. For modelling, resulting differences among the maps and their low accuracies have to be considered. The integration of an erroneous product into predictive models may have a devastating effect with respect to statements on future development scenarios and perspectives for an area. However, for many regions the overall relative quality of the existing products is not well known and has not been evaluated in depth. One important task is therefore to investigate the suitability of global products for further usage, as their accuracy may differ for different regions and landscape units depending on class composition.

In this study, six land-cover products of medium spatial resolution were compared for the trans-boundary Mekong Basin – an extensive watershed of one of the world's largest rivers. The advantage of the vast river basin is the fact that it covers nearly all climate zones and related environmental regimes – from high alpine to mountainous grasslands, from dense rainforest to sparse shrublands, and from agricultural production areas to wetland swamps, to name only some. The northwest–southeast-trending basin traverses near longitudinally parallel for over 2000 kilometres, and in its diversity is an optimal study area for product comparison. Furthermore, no study so far has focused on a comprehensive and analytical land-cover comparison for this region. Similarities and differences between the different global products are elucidated in detail, and the main sources of disagreement identified. Findings of this study can be used by the broader science community for identifying local hot-spot areas or thematic land-cover types that are particularly difficult for regional land-cover mapping in this area, and may also contribute to a better understanding of environmental model outputs that are based on global land-cover products.

2. Study area: the Mekong Basin

The Mekong Basin (see Figure 1) is a trans-boundary river basin, including territorial parts of six countries: namely China, Myanmar, Laos, Thailand, Cambodia, and Vietnam. It is home to a riparian population exceeding 72 million inhabitants. The Mekong is the ninth largest river in the world and with 4900 km length is also one of the longest rivers in the world. It originates in China's Qinghai Tibet plateau at an elevation of 5200 m, and at the Mekong Delta in Vietnam empties into the South China Sea. The Mekong Basin has a size of 795,000 km^2, and – for comparison – thus covers two times the size of all of Germany, and is even 100,000 km^2 larger than all of France. The basin – often called pan shaped – contains mountainous regions with steep narrow valleys in China, dominated by barren land, forests, shrublands as well as grasslands, and gets less mountainous and less incised in Laos and Thailand, where many tributary rivers join the Mekong, and where agricultural plateaus, such as the Thai Khorat Plateau, dominate. The basin then widens in the alluvial lowlands of Cambodia and southern Vietnam, defining the Mekong Delta, where rice farming, aquaculture, orchards, and coastal mangrove belts dominate the landscape (Campbell 2009; Gupta 2009; Kuenzer et al. 2012). The region has a high geopolitical relevance. Not only are Mekong riparian nations undergoing rapid socio-economic changes, but controversial developments such as hydropower projects along the main stem and the tributaries, the rapid expansion of cash crop plantations such as rubber and cashew, as well as challenges of infrastructure development, land grabbing, forced migration, environmental pollution, and a loss of biodiversity and river-related food resources (protein) all have lifted the Mekong Basin into the focus of attention of many large research networks, donor agencies, stakeholders, and decision-makers. Sound information on the state of the environment of this large catchment is a crucial need.

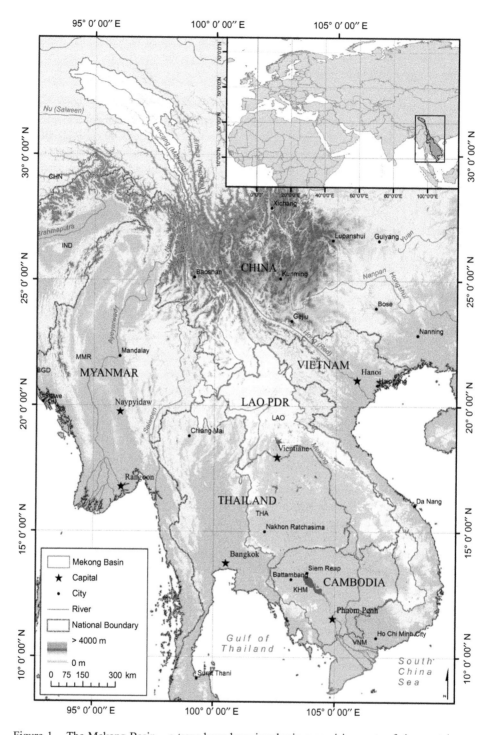

Figure 1.   The Mekong Basin – a trans-boundary river basin comprising parts of six countries.

3. Data product processing and harmonization procedure

For our study we compared six different land-cover maps stemming from four different classification approaches for the Mekong Basin and analysed their differences. These were as follows:

(1) UMD 1992/93 – University of Maryland Global Land Cover Product (DeFries et al. 1998; Hansen and Reed 2000; Hansen et al. 2000);
(2) GLC 2000 – Global Land Cover (Fritz and See 2005; Bartholomé and Belward 2005; Mayaux et al. 2006);
(3) GlobCover 2004–06 (Arino et al. 2007; Bicheron et al. 2008);
(4) GlobCover 2009;
(5) MODIS Land Cover MCD12Q1 2001; and
(6) MODIS Land Cover MCD12Q1 2009 (Friedl et al. 2010).

The product characteristics of these globally available data sets are presented in Table 1. It can be seen that the products were derived from different sensors, such as AVHRR, SPOT 4 VEGETATION, MERIS, and MODIS, represent land cover at different points in time, are characterized by a strongly varying number of land-cover classes, and also come at differing spatial resolutions, varying from 300 m to 1 km. Furthermore, it can be seen that the reported global area-weighted accuracy averages vary from 67.5% to 78.3%. However, no comparative conclusion on relative mapping performance can be derived from these accuracy statistics, owing to substantially differing classification legends and validation schemes (Herold et al. 2008). Generally, however, overall classification accuracies of global products can be expected to range between 65% and 75% (Fritz et al. 2011). It was ensured that data sets were geometrically corrected to each other, and a geographic coordinate system (lat/long) was chosen as the common reference projection.

To allow a comparison of the products, the land-cover classes of the individual products had to be translated to a common, harmonized legend. This was undertaken following Herold et al. (2008) and GOFC-GOLD Report No. 43. According to Table 1,

Table 1. Product characteristics of available global land-cover products.

Product characteristics	UMD	GLC 2000	GlobCover 2004/2009	MODIS Land Cover Type (MCD12Q1)
Sensor	AVHRR	SPOT 4	MERIS	MODIS
Time of data collection	April 1992–March 1993	Nov 1999–Dec 2000	Dec 2004–June 2006 Jan 2009–Dec 2009	2001–2011
Classification scheme	Simplified IGBP	FAO LCCS	UN LCCS	IGBP
	(14 classes)	(22 classes)	(22 classes)	(17 classes)
Spatial resolution	1 km	1 km	0.3 km	500 m
Accuracy	–	68.6% (Mayaux et al. 2006)	67.5% (Bontemps et al. 2011)	78.3%. (MODIS land-cover team, 2003)

Note: Based on Hansen and Reed (2000) (UMD), Fritz and See (2005) (GLC 2000), Bicheron et al. (2008) (GlobCover), and Friedl et al. (2010) (MODIS Land Cover).

the selected products are characterized by 14 up to 22 land-cover classes, which need to be translated into 13 classes determined by the LCCS classification scheme. This scheme is based on a general agreement of the UN Land Cover Classification System. It provides a valuable common land-cover language for building land-cover legends and translating and comparing existing legends (Herold et al. 2006). The LCCS defines classifiers rather than categories and thus standardizes terminology and the attributes used to define thematic classes in maps (Herold et al. 2008). In Table 2, the generalized global land-cover legend, with LCCS definitions and corresponding classes from the individual global legends, is listed. The translation of legends is based on Herold et al. (2008) (UMD, GLC 2000, and MODIS MCD12Q1), GOFC-GOLD Report No. 43 (UMD), and Bicheron et al. (2008) (GlobCover).

Translation to the common LCCS legend is mostly an unambiguous process. Ambiguities might occur in some selected cases, but can be resolved through clear definitions according to logical reasoning. Concerning the translation process the following additional product-specific remarks are relevant:

For the UMD products, class 6 (woodlands – defined as areas with a tree canopy cover of >40% and <60%, trees exceeding a height of 5 m) and class 7 (wooded grasslands/ shrublands – with a tree canopy cover of >10% and <40%, with trees exceeding 5 m in height) were translated to LCCS class 5 (mixed/other trees) because of their defined tree canopy cover and tree height and no further defined leaf-type characteristics. UMD class 12 (barren) does not distinguish between snow/ice and barren areas and is thus translated only to LCCS class 12 (barren) and no area is translated to LCCS class 11 (snow/ice).

For the GLC 2000 product, GLC classes 7 and 8 are translated to LCCS class 5 instead of 2. There is a contradiction in Herold et al. (2008), in which the author suggests in the written text to assign 7 and 8 to LCCS class 5, but in Table 2 it is assigned to LCCS class 2.

For GlobCover 2004–09, class 90 cannot be assigned explicitly. Fortunately in this case it is not necessary, because it does not occur in the Mekong Basin region at all.

Concerning the Land Cover Type product MCD12Q1, there are five different classifications available in the original product (IGBP, UMD, LAI/fPAR, NPP/BGC, and PFT). For the harmonization of the MCD12Q1 product to LCCS, the IGBP classification was chosen.

Based on the harmonized data, it is now possible to analyse the differences of the individual global products.

4. Results of harmonized product comparison

4.1. Results of a general qualitative and quantitative assessment

Figure 2 presents a visual comparison of the six harmonized land-cover maps for the Mekong Basin. The visual interpretation alone uncovers vast differences among the harmonized products.

The distribution of the GlobCover and especially the MODIS land-cover classes appears rather homogenous within the study area while those of the GLC2000 map and the UMD map show much more heterogeneous and fragmented patterns. While GlobCover 2004–06 and GlobCover 2009 are basically dominated by the classes 'Evergreen Broadleaf Trees' (2), 'Shrubs' (6), and 'Cultivated and Managed Vegetation/ Agriculture (incl. Mixtures)' (8), the MODIS maps as well as the UMD map are characterized by the classes 'Evergreen Broadleaf Trees' (2), 'Mixed/Other Trees' (5), 'Herbaceous Vegetation' (7), as well as 'Cultivated and managed Vegetation/Agriculture

Table 2. Generalized global land-cover legend with corresponding classes from individual global legends.

LCCS Class	Generalized class description	UMD Class	Class description	GLC 2000 Class	Class description	GlobCover 2004–09 Class	Class description	MODIS Land Cover Type Class	Class description
1	Evergreen needleleaf trees	1	Needleleaved evergreen trees	4	Tree cover, needleleaved, evergreen	70	Closed (>40%) needleleaved evergreen forest (>5 m)	1	Evergreen needleleaf forest
						(90)	Open (15–40%) needleleaved deciduous or evergreen forest (>5 m)		
2	Evergreen broadleaf trees	2	Broadleaved evergreen trees	1	Tree cover, broadleaved, evergreen	40	Closed to open (>15%) broadleaved evergreen and/or semi-deciduous forest (>5 m)	2	Evergreen broadleaf forest
3	Deciduous needleleaf trees	3	Needleleaved deciduous tree	5	Tree cover, needleleaved, deciduous	(90)	Open (15–40%) needleleaved deciduous or evergreen forest (>5 m)	3	Deciduous needleleaf forest
4	Deciduous broadleaf trees	4	Broadleaved deciduous tree	2	Tree cover, broadleaved, deciduous, closed	50	Closed (>40%) broadleaved deciduous forest (>5 m)	4	Deciduous broadleaf forest
				3	Tree cover, broadleaved, deciduous, open	60	Open (15–40%) broadleaved deciduous forest (>5 m)		
5	Mixed/other trees	5	Mixed Forests	6	Tree cover, mixed leaf type	100	Closed to open (>15%) mixed broadleaved and needleleavedforest (>5 m)	5	Mixed forest
		6	Woodlands	7	Tree cover, regularly flooded, fresh			8	Woody savannahs
				8	Tree cover, regularly flooded, saline				
				9	Mosaic: tree cover/other natural vegetation	160	Closed (>40%) broadleaved forest regularly flooded – Freshwater		

(Continued)

Table 2. (Continued).

LCCS		UMD		GLC 2000		GlobCover 2004-09		MODIS Land Cover Type	
Class	Generalized class description	Class	Class description	Class	Class description	Class	Class description	Class	Class description
		7	Wooded grasslands/ shrublands	10	Tree cover, burnt	170	Closed (>40%) broadleaved semi-deciduous and/or evergreen forest regularly flooded – saline water	9	Savannahs
6	Shrubs	8	Closed bushlands or shrublands	11	Shrub cover, closed-open, evergreen (with or without sparse tree layer)	130	Closed to open (>15%) shrubland (<5 m)	6	Closed shrublands
		9	Open shrublands	12	Shrub cover, closed-open, deciduous (with or without sparse tree layer)			7	Open shrublands
7	Herbaceous vegetation	10	Grasslands	13	Herbaceous Cover, closed-open	140	Closed to open (>15%) grassland	10	Grasslands
8	Cultivated and managed vegetation/ agriculture (incl. mixtures)	11	Croplands	16	Cultivated and managed areas	11	Post-flooding or irrigated croplands	12	Croplands
				17	Mosaic: cropland/tree cover/other natural vegetation	14	Rain-fed croplands		
				18	Mosaic: cropland/shrub and/or herbaceous cover	20	Mosaic cropland (50–70%)/ vegetation (grassland, shrubland, forest) (20-50%)	14	Cropland/natural vegetation mosaic

(Continued)

Table 2. (Continued).

LCCS Class	Generalized class description	UMD Class	Class description	GLC 2000 Class	Class description	GlobCover 2004-09 Class	Class description	MODIS Class	MODIS Land Cover Type Class description
						30	Mosaic vegetation (grassland, shrubland, forest) (50–70%)/ cropland (20–50%)		
9	Other shrub/ herbaceous vegetation	–	–	15	Regularly flooded shrub and/or herbaceous cover	110	Mosaic forest/shrubland (50–70%)/forest/shrubland (20–50%)	11	Permanent wetlands
						120	Mosaic grassland (50–70%)/ forest/shrubland (20–50%)		
						180	Closed to open (>15%) grassland or woody vegetation on regularly flooded or waterlogged soil – fresh, brackish, or saline water		
10	Urban/ built-up	13	Urban and built-up	22	Artificial surfaces and associated areas	190	Artificial surfaces and associated areas (urban areas >50%)	13	Urban and built-up
11	Snow and Ice	12	Barren	21	Snow and ice (natural & artificial)	220	Permanent snow and ice	15	Snow and ice
12	Barren			14	Sparse herbaceous or sparse shrub cover	150	Sparse (>15%) vegetation (woody vegetation, shrubs, grassland)	16	Barren or sparsely vegetated
				19	Bare areas	200	Bare areas		
13	Open water	0	Waterbodies	20	Waterbodies (natural & artificial areas)	210	Waterbodies	0	Water

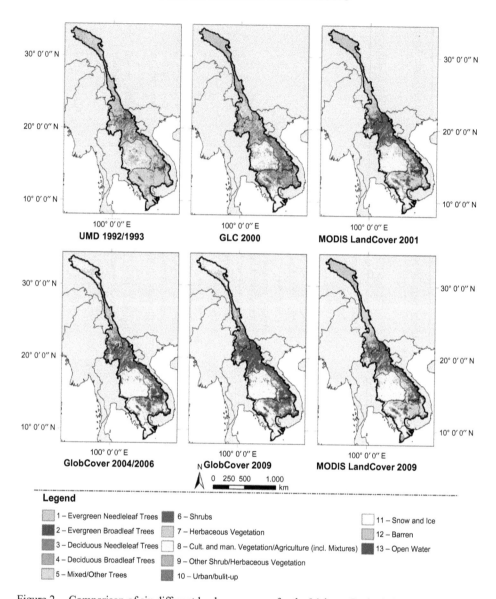

Figure 2.   Comparison of six different land-cover maps for the Mekong Basin (LCCS harmonized).

(incl. Mixtures)' (8). The similarities of the MODIS and UMD maps might be caused by the IGBP input-classification, which both products have in common.

Two classes underline this in an accentuated way: the strikingly different manifestation of the class 'Herbaceous Vegetation' (LCCS 7) in the very northern panhandle of the basin. This class is very poorly represented in the UMD as well as in the GlobCover products. LCCS class 8 (cultivated agricultural area) seems to cover similar extents in the MODIS, as well as the GlobCover products, but has a much smaller extent in the GLC 2000 products, and seems nearly non-existent in the UMD product. Nevertheless, the general patterns of land cover displayed in the maps are relatively similar: regions of intense agricultural usage are easily distinguishable from regions of major woodland coverage.

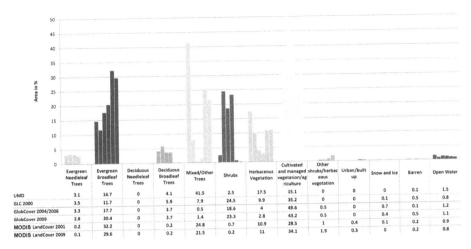

	Evergreen Needleleaf Trees	Evergreen Broadleaf Trees	Deciduous Needleleaf Trees	Deciduous Broadleaf Trees	Mixed/Other Trees	Shrubs	Herbaceous Vegetation	Cultivated and managed vegetation/ag riculture	Other shrubs/herbac eous vegetation	Urban/built up	Snow and Ice	Barren	Open Water
UMD	3.1	14.7	0	4.1	41.5	2.5	17.5	15.1	0	0	0	0.1	1.5
GLC 2000	3.5	11.7	0	5.9	7.9	24.5	9.9	35.2	0	0	0.1	0.5	0.8
GlobCover 2004/2006	3.3	17.7	0	3.7	0.5	18.6	4	49.6	0.5	0	0.7	0.1	1.2
GlobCover 2009	2.8	20.4	0	3.7	1.4	23.3	2.8	43.2	0.5	0	0.4	0.5	1.1
MODIS LandCover 2001	0.2	32.2	0	0.2	24.8	0.7	10.9	28.5	1	0.4	0.1	0.2	0.9
MODIS LandCover 2009	0.1	29.6	0	0.2	21.5	0.2	11	34.1	1.9	0.3	0	0.2	0.8

Figure 3. Area coverage in per product-related LCCS land-cover class (in percentage) for the Mekong Basin.

When quantitatively analysing the differences, as elucidated in Figure 3, the following conclusions can be derived. First of all, LCCS classes 9–13 ('Other Shrubs', 'Urban/Built Up', 'Snow and Ice', 'Barren', and 'Open Water') can mostly be ignored for the comparison. Each of them occur at very small percentages within the basin, and especially the waterbodies class cannot really be assessed in an environment characterized by alternating rainy and dry seasons. But actually, this latter class does not vary too strongly. Waterbodies are visually detectable in all land-cover maps, especially the Tonle Sap Lake in Cambodia can be recognized. Apart from this, the UMD map accounts for the biggest part of 'Open Water' in the Mekong Basin, also depicting smaller features such as lakes or rivers, in particular the Mekong itself.

However, the other classes are considered suitable for further investigation. While the LCCS classes 'Evergreen Needleleaf Trees' and 'Deciduous Broadleaf Trees' are more or less evenly distributed in four of the six compared data sets (UMD, GLC 2000, GlobCover) – here ranging between 2.8–3.5% and 3.7–5.9%, respectively – they are more or less non-existent in the harmonized MODIS product, where they reach a maximum coverage of 0.2%. The classes 'Evergreen Broadleaf Trees', 'Mixed/Other Trees', 'Shrubs', and 'Herbaceous Vegetation' are extremely variable and contributions within the basin vary in the range 11.7–32.2%, 0.5–41.5%, 0.2–24.5%, and 2.8–17.5%, respectively. Even the class depicting 'Cultivated agricultural land' shows a strong variability, with areal catchment proportions ranging from 15.1% to 49.6%.

A temporal assessment and analysis can be performed based on Figure 4. We can see that the GlobCover 2009 as well as the MODIS land-cover product from 2009 was generated from data of exactly the same year: the former from MERIS data and the latter from MODIS data. However, although they are from the same year, the differences between the two products are very obvious. GlobGover 2009 displays a much larger areal coverage with cultivated agricultural land compared with the MODIS products. The shrubland areas distinguished in GlobCover 2009 do not exist in the MODOS 2009 product. In the latter, the herbaceous class is more accentuated. However, a closer look at all forest-cover classes reveals that the products are far from similar or comparable, even when derived for the same year. Although the GlobCover 2009 product assigns slightly more than one quarter of the basin to tree-cover (forest) classes, the MODIS 2009

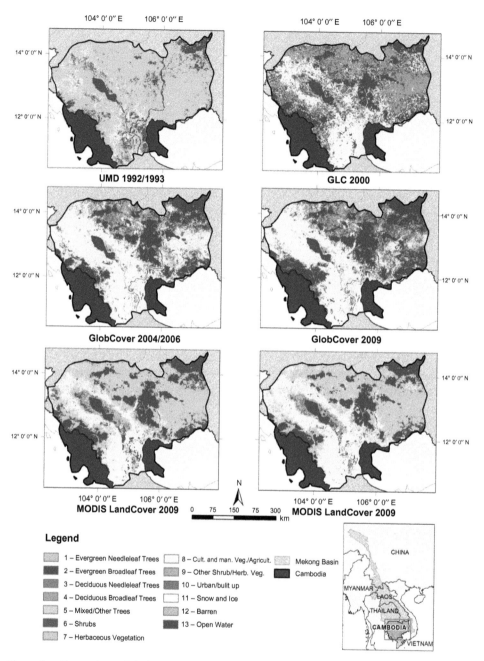

Figure 4. Comparison of six land-cover maps for the Cambodian part of the Mekong Basin (LCCS harmonized).

Land Cover product assigns over 50% to these classes. What is defined as 'Mixed/Other Trees' with the MODIS product is the 'Shrubs' class in the GlobCover product. However, the 'Herbaceous' and the 'Cultivated' class jointly sum up to about 45% areal coverage of the basin in both cases. Thus, although very broad general patterns seem to be captured by both classifications, significant differences (definitions of shrubland vs. tree cover) prevail, which will definitely impact modelling applications. It is furthermore important to note that differences between the individual products cannot be interpreted as natural or human-driven land-use changes, since the products vary in production algorithm, data resolution, and season of data acquisition, amongst others.

Figure 4 presents the six harmonized products at greater detail for the Cambodian part of the Mekong Basin. All the above-described phenomena become even more obvious here.

The agreement of waterbodies is very good – especially for the Tonle Sap Lake, whereas the Mekong is best visible in the UMD93 product, which also depicts the highest overall values for the water class. The class comprising evergreen broadleaf forest shows high agreements amongst GLC 2000, the GlobCover product, and the MODIS products, but strong disagreements with the UMD product. The expansion of the agricultural area (which did occur as can be confirmed by Landsat data, country statistics, and other studies) can be impressively visualized comparing the MODIS 2001 with the MODIS 2009 product. Here – as one of the few exceptions – global products can be used also to assess land-surface dynamics, as long as they were derived with the same algorithm and are from the same 'product family' (in this case both MODIS). Similar to the overall basin, we can see that maps that assign large areas to class 6 (Shrubs) are nearly free of class 5 (Mixed/Other Trees) and vice versa. This can also be observed when calculating the cross-tabulation between the products later on. In the six land-cover maps urban areas can only be seen in the MODIS products. Here Phnom Penh, the capitol of Cambodia, can be recognized southeast of the Tonle Sap Lake. In the other maps this city was not mapped.

4.2. Results of the comparison of individual classes and merged classes

The following Figures 5–7 present the assessment of individual LC classes. We combined all tree-cover classes (LCCS classes 1, 2, 3, 4, and 5) to elucidate which parts of the Mekong Basin are considered to be forested areas. Figure 5 shows that the largest forest-cover extents can be found in the UMD 1992/1993 product (at this stage we can only speculate whether this is related to the early date of this product and reflects a more untouched state of the basin, which nowadays undergoes much change due to deforestation), whereas GLC 2000 and the GlobCover products show a much lower forest-cover extent – also in comparison with the MODIS product. The large differences in forest-cover distribution can be explained by the rather large quality differences of the respective forests in this area. Particularly in the highlands of northern Laos, primary and secondary broadleaved evergreen forest formation regions occur. In the lower valleys, however, natural forests are heavily degraded and often converted to evergreen and deciduous woodlands and shrublands, interspersed by small patches of cropland and forest cuts used for less-intensive cultivation. Although the MODIS and UMD products incorporate these heavily degraded, semi-natural vegetation types in the forest classes, the GlobCover and GLC 2000 products frequently classified these areas as shrubland and vegetation/cropland mosaics (see Figure 6). As a result forest cover constitutes around 52–61% of the Basin according to MODIS or UMD Land Cover, whereas GlobCover and GLC 2000 only estimate approximately 22–16% of the Basin's area being forested. On the other hand GlobCover and GLC 2000 indicate around 19–25% of the Basin as shrublands, whereas the latter LC class hardly occurs in the MODIS or UMD land-cover maps. Forest-cover extent in the MODIS 2009 product is diminished compared to the

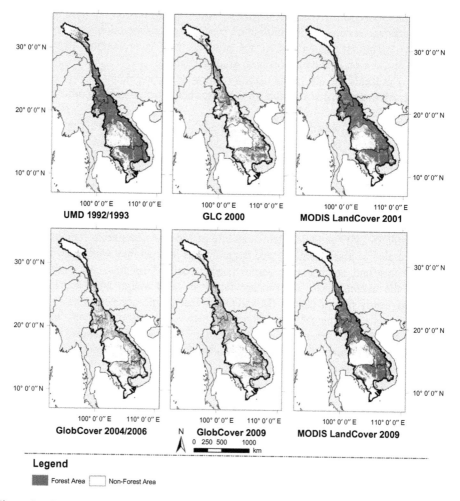

Figure 5.   Comparison of forest area coverage in the Mekong Basin.

MODIS 2001 product – as these two maps were derived via the same algorithm here a temporal development (deforestation) might be responsible for the change.

Moreover, when comparing cultivated and non-cultivated land (see Figure 7), large differences between the individual products occur. The lowest basin coverage of culti-vated land (15.1% of the basin) is displayed in the UMD product, whereas all other products depict a much larger cultivated area. The GLC 2000 product indicates that 35.2% of the basin is cultivated, whereas the two GlobCover products of 2004–2006 and 2009 indicate 49.6% and 43.2%, respectively. In the MODIS product of 2001 28.5% of the basin is indicated to be cultivated, whereas in 2009 it is 34.1%. Similar among all products is that the Khorat Plateau of north-eastern Thailand in the lower centre of the basin stands out as the main agricultural area within the basin in all products. In the 2000 and 2001 GLC and MODIS products, the plains of Cambodia and the Mekong Delta were mapped as prominent cultivated areas. However, in the 1990s the Mekong Delta was an intensive agricultural area (Renaud and Kuenzer 2012), completely dominated by rice farming, and the UMD product thus clearly underestimates the cultivated area here. Although we want to underline that we are well aware that these

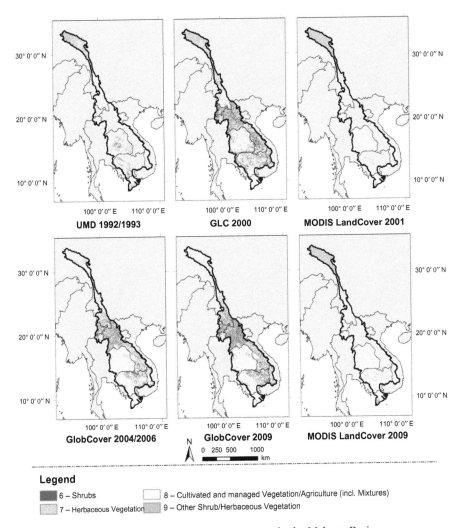

Figure 6.   Comparison of non-forest vegetation coverage in the Mekong Basin.

global products (owing to their relatively low local or even regional accuracy) cannot be used for any kind of temporal change detection, some inconsistencies appear striking while some statements on developments within classes can still be derived. Knowing that the cultivated area in the Mekong Basin has expanded on the coast due to forest clear cutting and other land transformation, it seems unlikely that in the GlobCover 2009 product the cultivated area's contribution to the basin is, at 43.2%, over 6% lower than in the same product of 2004–2006. Here the tendency displayed within the MODIS product, where the cultivated area covers 28.5% in 2001 and 34.1% in 2009, appears to be more reasonable.

The class with the lowest representation within the analysed products is the urban class. Experiences from recent land-cover classification approaches have shown that the mapping of urban areas based solely on low-resolution data have proved to be a rather challenging task. Urban core areas are often confused with barren land whereas suburban regions are easily misinterpreted as being natural vegetation (Friedl et al.

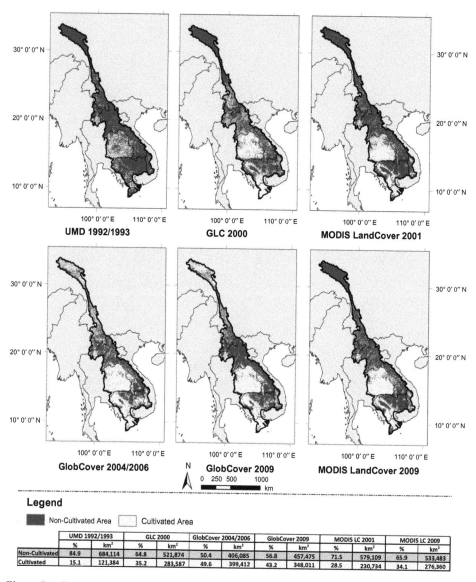

Figure 7. Comparison of cultivated area coverage in the Mekong Basin.

2010). These difficulties are clearly reflected in the strong differences of urban area coverage that can be observed here (Figure 8). While urban areas in the UMD, GLC 2000, and GlobCover products are at very low levels, between 0.01 and 0.05%, the proportion of the urban class, significantly larger in the MODIS products at 0.35%, is up to 40 times higher than in the other maps. These large differences can be explained by the fact that the urban class in the MODIS products originates from an external product that solely focuses on the extraction of urban areas (A. Schneider, Friedl, and Potere 2010). In this respect, accuracy for the urban class, which generally suffers from errors of omission, can be expected to be far higher in the MODIS product.

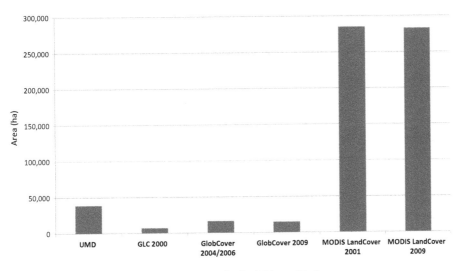

Figure 8. Comparison of urban area coverage in the Mekong Basin.

4.3. Results of the quantitative analyses of cross-product coincidence

In the following we present the results of the quantification of differences between the individual land-cover products via cross-tabulation. In an ideal case the values in the centre diagonal of the matrix should be 100%. Table 3 depicts the cross-tabulation matrix between the LCCS harmonized GlobCover 2009 and the MODIS Land Cover 2009 product, and in Table 4 the cross-tabulation matrix between the two GlobCover products is shown. Although the GlobCover 2009 and the MODIS 2009 products represent the same year, whereas the two GlobCover products of 2004–2006 and 2009 include nearly half a decade of temporal gap, it is obvious that differences between temporally similar products (GlobCover 2009–MODIS Land Cover 2009/GLC2000–MODIS Land Cover 2001) are much bigger than between two maps of the same product, which differ only in time (GlobCover 2004–06–GlobCover 2009/MODIS 2001–MODIS 2009).

Table 5 shows the overall agreement between all combinations of the six land-cover maps presented in this study. Although not all comparisons are reasonable because of large temporal gaps, where land-cover dynamics over time might confuse the results (UMD93 to GlobCover 2009), it can be shown that related products (GlobCover 2004–06 and GlobCover 2009 as well as MODIS 2001 and 2009) account for much greater agreements than unrelated products of similar data assessment times (GlobCover 2009 to MODIS 2009/GLC2000 to MODIS 2001) (Figure 9). However, it is important to point out that additional disagreement between unrelated map products might also arise from their different sensor resolutions. For the analysed maps, the size of the sensor's projected instantaneous field of view varies by a maximum factor of 10, which particularly effects areas where multiple land-cover types occur within the extent of a pixel (mixed pixels). Although a courser resolution product identifies, e.g., alternating patches of forests and grassland within a pixel uniformly as woodland or as a mosaic class, finer resolution maps may be able to differentiate these alternating patches into two land-cover categories. Although both products correctly map these patterns according to their respective resolutions, the courser map includes a higher percentage of mixed pixels that give rise to an overall underestimation of non-dominant and fragmented land-cover types (Nelson and Holben 1986; Braswell et al. 2003; Leinenkugel et al. 2014).

Table 3. Cross-tabulation matrix between GlobCover 2009 (LCCS) and MODIS Land Cover 2009 (LCCS), from the perspective of GlobCover 2009, displayed as percentages. LCCS classes of GlobCover09 are displayed in columns; the portions of the corresponding MODIS Land Cover LCCS classes are displayed in rows. Major classes of both products are highlighted in grey. Values exceeding 30% on the centre diagonal are highlighted in green; others in red.

		GlobCover 2009												
		Class01 (%)	Class02 (%)	Class03 (%)	Class04 (%)	Class05 (%)	Class06 (%)	Class07 (%)	Class08 (%)	Class09 (%)	Class10 (%)	Class11 (%)	Class12 (%)	Class13 (%)
MODIS 2009	Class01	**2.9**	0.01	0.0	0.0	0.3	0.00	0.1	0.02	5.6	0.0	0.3	0.0	0.1
	Class02	20.6	**76.3**	0.0	6.0	11.6	53.5	0.3	1.5	3.1	0.8	0.0	0.0	4.6
	Class03	0.0	0.0	**0.0**	0.0	0.0	0.00	0.0	0.00	0.0	0.0	0.0	0.0	0.0
	Class04	0.1	0.2	0.0	**0.7**	0.4	0.4	0.0	0.1	0.1	0.0	0.0	0.0	0.1
	Class05	47.9	15.7	0.0	70.6	**50.1**	35.2	0.8	12.4	38.4	6.5	0.9	0.1	4.3
	Class06	1.1	0.0	0.0	0.1	0.5	**0.0**	0.5	0.1	2.8	0.4	0.1	0.0	0.9
	Class07	22.8	0.1	0.0	1.8	11.1	0.4	**82.8**	15.4	43.8	5.5	91.9	98.3	2.8
	Class08	1.4	7.5	0.0	20.2	14.6	9.9	8.5	**67.0**	1.6	44.0	0.1	0.9	15.5
	Class09	2.5	0.2	0.0	0.4	10.9	0.4	4.6	2.5	**3.9**	2.3	0.0	0.0	14.3
	Class10	0.0	0.0	0.0	0.1	0.1	0.0	0.2	0.7	0.0	**39.6**	0.0	0.1	0.8
	Class11	0.0	0.0	0.0	0.0	0.0	0.0	0.0	0.0	0.1	0.0	**6.1**	0.4	0.1
	Class12	0.0	0.0	0.0	0.0	0.0	0.0	1.4	0.1	0.1	0.9	0.2	**0.1**	7.3
	Class13	0.6	0.0	0.0	0.1	0.4	0.1	0.8	0.2	0.6	0.1	0.4	0.1	**49.3**
	Total:	100.0	100.0	0.0	100.0	100.0	100.0	100.0	100.0	100.0	100.0	100.0	100.0	100.0

Table 4. Cross-tabulation matrix between GlobCover 2004–06 (LCCS) and GlobCover 2009 (LCCS), displayed as percentages, from the perspective of GlobCover 2004–06. Major classes of both classifications are highlighted in green; others in grey. Values exceeding 30% on the centre diagonal are highlighted in red.

		GlobCover 2004–06												
		Class01 (%)	Class02 (%)	Class03 (%)	Class04 (%)	Class05 (%)	Class06 (%)	Class07 (%)	Clas08 (%)	Class09 (%)	Class10 (%)	Class11 (%)	Class12 (%)	Class13 (%)
GlobCover 2009	Class0	51.2	1.2	0.0	0.6	3.2	0.2	1.1	1.3	16.6	0.2	0.5	0.4	0.8
	Class02	21.6	77.8	0.0	8.3	3.8	11.0	0.0	7.1	0.3	1.4	0.0	0.0	0.7
	Class03	0.0	0.0	0.0	0.0	0.0	0.0	0.0	0.0	0.0	0.0	0.0	0.0	0.0
	Class04	0.5	0.2	0.0	63.5	4.5	2.8	0.0	1.5	0.2	0.4	0.0	0.0	0.4
	Class05	12.4	1.7	0.0	0.8	59.4	0.4	0.1	0.5	2.9	0.1	0.1	0.1	1.6
	Class06	6.4	18.6	0.0	17.4	11.6	83.9	0.1	6.8	1.6	0.7	11.5	22.0	2.1
	Class07	0.6	0.0	0.0	0.0	0.8	0.1	49.5	1.3	2.7	0.2	9.5	11.2	10.5
	Class08	4.5	0.5	0.0	9.3	16.3	1.5	44.9	81.2	7.0	15.8	0.3	0.1	0.1
	Class09	2.6	0.0	0.0	0.0	0.2	0.0	0.2	0.1	68.3	0.0	0.0	0.0	0.1
	Class10	0.0	0.0	0.0	0.0	0.0	0.0	0.0	0.0	0.0	78.2	34.7	3.2	0.0
	Class11	0.0	0.0	0.0	0.0	0.0	0.0	2.0	0.0	0.2	0.0	43.4	63.0	0.0
	Class12	0.0	0.0	0.0	0.0	0.0	0.0	2.1	0.0	0.1	0.1	0.0	0.0	0.0
	Class13	0.2	0.0	0.0	0.1	0.3	0.1	0.1	0.2	0.3	3.0	0.0	0.0	83.4
	Total:	100.0	100.0	0.0	100.0	100.0	100.0	100.0	100.0	100.0	100.0	100.0	100.0	100.0

LCCS class	Name
1	Evergreen needleleaf trees
2	Evergreen broadleaf trees
3	Deciduous needleleaf trees
4	Deciduous broadleaf trees
5	Mixed/other trees

LCCS class	Name
6	Shrubs
7	Herbaceous vegetation
8	Cultivated and managed vegetation/agriculture
9	Other shrubs/herbaceous vegetation
10	Urban/built-up

LCCS class	Name
11	Snow and ice
12	Barren
13	Open water

Table 5. Overall classification agreements (LCCS harmonized) between the presented land-cover maps.

Combination	Agreement (%)	km²
MODIS 2001 – GLC 2000	**41.5**	**334.0**
MODIS 2001 – GlobCover 2004–06	46.1	371.60
MODIS 2001 – GlobCover 2009	45.0	362.4
MODIS 2001 – MODIS 2009	78.9	635.5
MODIS 2001 – UMD 93	48.0	386.3
MODIS 2009 – GLC 2000	43.9	353.2
MODIS 2009 – GlobCover 2004–06	50.0	402.5
MODIS 2009 – GlobCover 2009	**48.0**	**388.7**
MODIS 2009 – UMD 93	47.3	381.3
GLC 2000 – GlobCover 2004–06	46.7	376.4
GLC 2000 – UMD 93	26.1	209.9
GLC 2000 – GlobCover 2009	46.1	371.2
GlobCover 2004–06 – GlobCover 2009	77.7	625.8
GlobCover 2004–06 – UMD 93	25.1	202.1
GlobCover 2009 – UMD 93	24.3	195.4

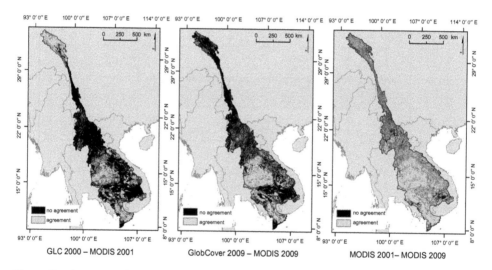

Figure 9. Agreement between different land-cover products of similar assessment times and similar land-cover products of different assessment time points.

Figure 10 elucidates the number of coincidences between the six individual products, as well as between the four fully independent products, respectively. Agreements mainly occur in the northern panhandle of the basin due to the coincidence in mapping herbaceous vegetation (LCCS class 7), in the cultivated lowlands of northeast Thailand (Khorat Plateau), in cultivated areas of western Cambodia and the Mekong Delta Region of Vietnam, as well as in regions of evergreen broadleaf forest (especially the Annamite Range – northeast Cambodia, eastern Laos – western Vietnam). Highest agreements occur for waterbodies – here especially Tonle Sap Lake. The largest disagreements occur in the vicinity of Tonle Sap Lake, which is due to the high fluctuations in the water levels of the lake between the dry and rainy seasons (Kuenzer 2013), in the south-western tip of the Mekong Delta, where also aquaculture and flooding-related water-level variations might hamper land-

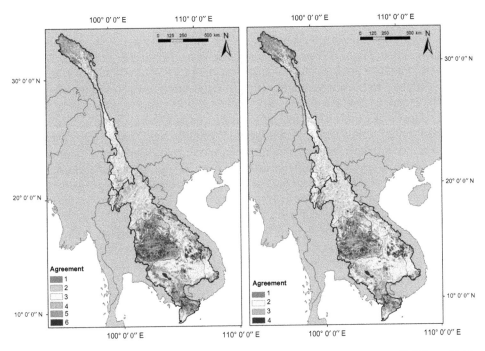

Figure 10. Comparison of land-cover products (UMD93, GLC2000, GlobCover04-06, GlobCover09, MODIS2001, and MODIS2009), regarding the agreement of assigned LCCS classes. Pixels with agreement of one to six classifications are shown, regardless of the assigned LCCS land-cover class (left). Comparison of land-cover products (UMD93, GLC2000, GlobCover20004–06, and MODIS2001) regarding agreement of assigned LCCS classes. Only unrelated classification products are considered in the comparison (right).

cover classification (Kuenzer et al. 2013), as well as in the highly mixed land-cover regions of southern Kunming, China, Laos, and eastern Cambodia and Vietnam.

Some disagreements between the maps, however, may also arise from real land-cover changes on the ground. The rapid socio-economic growth within the region has substantially increased pressure on the basin's natural resources, which are increasingly subject to over-exploitation and environmental degradation. Particularly, the forested areas in the region face high pressure through the expansion of farmland and industrial cash crop plantations. According to the national statistics from the Food and Agriculture Organization of the United Nations Global Forest Resources Assessments (FRA FAO 2010), it can be assumed that forested areas in Laos and Cambodia decreased by an annual rate of 1.26% and 0.47% throughout the last decade, respectively. The forests in the economically more powerful countries of Thailand and Vietnam, in contrast, show close to no losses, at a rate of 0.02% and increases in forested areas at 1.78% per year, respectively. Nevertheless, it can be expected that disagreement between the maps attributable to real land-cover modifications is rather small in relation to the amount of total disagreement between the individual products, as already demonstrated in Figure 10.

5. Conclusion

Based on our findings comparing the global land-cover classification products, we derive the following region-specific as well as general conclusions.

5.1. Region-specific conclusion

For the large trans-boundary Mekong River basin, the comparison of six globally available land-cover products reveals very large differences. Although GlobCover 2004–06 and GlobCover 2009 are basically dominated by the LCCS classes 'Evergreen Broadleaf Trees', 'Shrubs', and 'Cultivated and Managed Vegetation/Agriculture (incl. Mixtures)', the MODIS-derived land-cover maps as well as the UMD map are characterized by the LCCS classes 'Evergreen Broadleaf Trees', 'Mixed/Other Trees', 'Herbaceous Vegetation', and 'Cultivated and Managed Vegetation/Agriculture (incl. Mixtures)'. Obvious is the strikingly different manifestation of the class 'Herbaceous Vegetation' in the very northern panhandle of the basin. This class is very weakly accentuated in the UMD and the GlobCover products. The LCCS class and 'Cultivated and Managed Vegetation/Agriculture (incl. Mixtures)' seems to cover similar extents in the MODIS and the GlobCover products, but has a much smaller extent in the GLC 2000 products, and seems nearly non-existent in the UMD product. In the LCCS classes 'Other Shrubs', 'Urban/Built Up', 'Snow and Ice', 'Barren', and 'Open Water' can mostly be omitted for comparison. Each of them only occurs at very small percentage within the basin. Concerning all other classes, the 'Evergreen Needleleaf Trees' and 'Deciduous Broadleaf Trees' are more or less evenly distributed in four of the six compared data sets (UMD, GLC 2000, and GlobCover) – here ranging between 2.8–3.5% and 3.7–5.9%, respectively – while being nearly non-existent in the harmonized MODIS product, where they reach a maximum coverage of 0.2%. Extremely variable are the classes 'Evergreen Broadleaf Trees', 'Mixed/Other Trees', 'Shrubs', and 'Herbaceous Vegetation'. Here contributions within the basin vary in the range 11.7–32.2%, 0.5–41.5%, 0.2–24.5%, and 2.8–17.5%, respectively. Even the class depicting 'Cultivated and Managed Vegetation/Agriculture (incl. Mixtures)' shows strong variability, with contributions ranging from 15.1% to 49.6%. Forest cover constitutes around 52–61% of the Basin according to MODIS or UMD Land Cover, whereas GlobCover and GLC 2000 only estimate approximately 22–16% of the Basin's area being forested. On the other hand GlobCover and GLC 2000 indicate around 19–25% of the Basin as shrublands, while the latter LC class hardly occurs in the MODIS or UMD land-cover maps. Although urban areas in the UMD, GLC 2000, and GlobCover product remain at a very low level between 0.01 and 0.05%, the proportion of the urban class is, at 0.35%, significantly larger in the MODIS products showing proportions being up to 40 times higher than in the other maps. These large differences can be explained by the fact that the urban class in the MODIS products originates from an external product that solely focuses on the extraction of urban areas.

Overall, product differences are enormous and coincidence is low, as revealed by cross-tabulation. Although coincidence between the MODIS 2001 and the MODIS 2009 products is nearly 79%, and between the two GlobCover products it is about 78%, coincidence between, e.g., the GLC 2000 and the MODIS 2001 product it only reaches 41%, and coincidence between the MODIS 2009 and the GlobCover 2009 products only reaches 48%. In the spatial context, the 'easiest to map' areas in the Mekong Basin appear to be the Khorat Plateau in north-eastern Thailand, as well as the agricultural plains of Cambodia and the Mekong Delta. Difficult to map seem to be the highly fragmented and highly dynamic environments in the vicinity of Tonle Sap Lake, as well as the mosaic of classes in southern Yunnan, as well as north and central Laos.

None of the investigated products should be integrated into Mekong Basin-related modelling activities without high precaution and clear knowledge about the strengths and weaknesses of the individual land-cover products and the comparison among them.

5.2. *General conclusion*

The Land Cover Classification System, LCCS, provides a good, standardized, and widely accepted means to harmonize large land-cover data sets with varying legends. It therefore enables the comparison of land-cover products, as well as the qualitative and quantitative assessment of product differences. However, at the same time it must be kept in mind that harmonized results are still hampered by the fact that classification algorithms differ for the individual products, and that classes with identical names might have different class boundaries. Furthermore, global land-cover products usually differ in their year of origin. Therefore – even after harmonization – differences elucidated based on the comparison might not only be attributable to the classification algorithm used or the class definition applied, but also to actual land-cover change, which occurs over the course of time. However, we have shown that products of similar temporal characteristics (derived for the same year) show strong disagreements when derived from differing classification approaches, whereas related products show strong agreements despite temporal lags. On the other hand, however, even strong agreement among maps does not necessarily guarantee accuracy. As differences between the analysed land-cover products are enormous, attention has to be paid when using these classifications as input for further modelling and environmental assessments. Results of modelling or related studies will always depend on the chosen input land-cover product.

Global land-cover products are of vital importance to the environmental science community with respect to any kind of global-scale analyses. On smaller scales, however, these products have been shown to be affected by local and regional biases, thereby making them rather inappropriate for regional-specific analyses. The differences in land-cover distribution between the respective products arise from the difficulties of generalizing the heterogeneity in land-cover characteristics within a couple of classes that should be valid for the entire planet, irrespective of climate, topography, or latitude. Regional-specific approaches for land-cover mapping, such as the Mekong LC 2010 product (Leinenkugel et al. 2013) for the Mekong Area, which has recently been published, might have the capacity to address these heterogeneities through the inclusion of regional-specific classification schemes, regional-tuned data processing, or through the consideration of auxiliary geoinformation and local expert knowledge. However, the generation of regional-specific products is laborious and costly and often not the main focus of the respective research topic. Instead, land-cover information is often considered as pre-existing, regardless of any quality issues. In this respect, it is important to note that the use of global-scale land-cover products for environmental modelling on regional scales will inevitably result in large differences in model results dependent on the respective land-cover product used.

In general, much more attention needs to be paid to the origin, advantages, and shortcomings of globally or regionally available large-scale information products derived from remote-sensing data. Inconsistencies and knowledge gaps can be bridged by studies exploring such data inconsistencies, by improved documentation of available land-cover products, by very careful selection of the most suitable product, and by an increased effort of the Earth Observation community to provide high-quality products including quality layers, especially for heterogeneous areas.

Acknowledgement

The authors thank Paul Schumacher for support with data download and initial GIS analyses.

Funding

The authors thank the German Federal Ministry of Education and Research, BMBF, for funding the German–Vietnamese WISDOM project (www.wisdom.eoc.dlr.de).

References

Arino, O., P. Bicheron, F. Achard, J. Latham, R. Witt, and J. Weber. 2008. "GLOBCOVER: The Most Detailed Portrait of Earth." *ESA Bulletin* 136: 24–31.

Baker, J. M., and T. J. Griffis. 2009. "Evaluating the Potential Use of Winter Cover Crops in Corn–Soybean Systems for Sustainable Co-Production of Food and Fuel." *Agricultural and Forest Meteorology* 149 (12): 2120–2132. doi:http://dx.doi.org/10.1016/j.agrformet.2009.05.017.

Baker, T. J., and S. N. Miller. 2013. "Using the Soil and Water Assessment Tool (SWAT) to Assess Land Use Impact on Water Resources in an East African Watershed." *Journal of Hydrology* 486: 100–111.

Bartholomé, E., and A. S. Belward. 2005. "GLC2000: A New Approach to Global Land Cover Mapping from Earth Observation Data." *International Journal of Remote Sensing* 26 (9): 1959–1977. doi:10.1080/01431160412331291297.

Biazin, B., and G. Sterk. 2013. "Drought Vulnerability Drives Land-Use and Land Cover Changes in the Rift Valley Dry Lands of Ethiopia." *Agriculture, Ecosystems & Environment* 164: 100–113. doi:10.1016/j.agee.2012.09.012.

Bicheron, P., P. Defourny, C. Brockmann, L. Schouten, C. Vancutsem, M. Huc, S. Bontemps, M. Leroy, F. Achard, M. Herold, F. Ranera, and O. Arino. 2008. "GLOBCOVER: Product Description and Validation Report." Accessed May 2009. http://postel.mediasfrance.org/en/DOWNLOAD/Documents/#globcover

Bontemps, S., P. Defourny, E. Van Bogaert, O. Arino, V. Kalagirou, and J. Ramos Perez. 2011. *Globcover 2009 Product–Description and Validation Report V. 2.2.* Louvain-La-Neuve: Université Catholique de Louvain (UCL) & Paris: European Space Agency (ESA).

Braswell, B. H., S. C. Hagen, S. E. Frolking, and W. A. Salas. 2003. "A Multivariable Approach for Mapping Sub-Pixel Land Cover Distributions Using MISR and MODIS: Application in the Brazilian Amazon Region." *Remote Sensing of Environment* 87 (2–3): 243–256.

Brown, G. 2013. "The Relationship Between Social Values for Ecosystem Services and Global Land Cover: An Empirical Analysis." *Ecosystem Services.* doi:http://dx.doi.org/10.1016/j.ecoser.2013.06.004.

Campbell, I. C. 2009. *The Mekong–Biophysical Environment of an International River Basin.* New York: Elsevier.

Colditz, R. R., G. L. Saldaña, P. Maeda, J. A. Espinoza, C. M. Tovar, A. V. Hernández, C. Z. Benítez, I. C. López, and R. Ressl. 2012. "Generation and Analysis of the 2005 Land Cover Map for Mexico Using 250 M MODIS Data." *Remote Sensing of Environment* 123: 541–552.

Dong, J., X. Xiao, S. Sheldon, C. Biradar, N. D. Duong, and M. Hazarika. 2012. "A Comparison of Forest Cover Maps in Mainland Southeast Asia from Multiple Sources: PALSAR, MERIS, MODIS and FRA." *Remote Sensing of Environment* 127: 60–73. doi:10.1016/j.rse.2012.08.022.

Dunjó, G., G. Pardini, and M. Gispert. 2004. "The Role of Land Use–Land Cover on Runoff Generation and Sediment Yield at a Microplot Scale, in a Small Mediterranean Catchment." *Journal of Arid Environments* 57 (2): 239–256. doi:http://dx.doi.org/10.1016/S0140-1963(03)00097-1.

Eisfelder, C., C. Kuenzer, and S. Dech. 2011. "Derivation of Biomass Information for Semi-Arid Areas Using Remote-Sensing Data." *International Journal of Remote Sensing* 33 (9): 2937–2984. doi:10.1080/01431161.2011.620034.

FRA FAO. 2010. *Global Forest Resources Assessment 2010.* Forestry Paper No 163. Rome: Food and Agriculture Organization of the United Nations.

Friedl, M. A., D. Sulla-Menashe, B. Tan, A. Schneider, N. Ramankutty, A. Sibley, and X. Huang. 2010. "MODIS Collection 5 Global Land Cover: Algorithm Refinements and Characterization of New Datasets." *Remote Sensing of Environment* 114 (1): 168–182. doi:10.1016/j.rse.2009.08.016.

Fritz, S., and L. See. 2005. "Comparison of Land Cover Maps Using Fuzzy Agreement." *International Journal of Geographical Information Science* 19: 787–807.

Fritz, S., L. See, I. McCallum, C. Schill, M. Obersteiner, M. van der Velde, H. Bottcher, P. Havlik, and F. Achard. 2011. "Highlighting Continued Uncertainty in Global Land Cover Maps for the User Community." *Environmental Research Letters* 6: 044005.

Fritz, S., L. See, and F. Rembold. 2010. "Comparison of Global and Regional Land Cover Maps with Statistical Information for the Agricultural Domain in Africa." *International Journal of Remote Sensing* 31: 2237–2256.

Gao, J., and Y. Liu. 2010. "Determination of Land Degradation Causes in Tongyu County, Northeast China via Land Cover Change Detection." *International Journal of Applied Earth Observation and Geoinformation* 12 (1): 9–16. doi:10.1016/j.jag.2009.08.003.

GEO. 2005. *Global Earth Observation System of Systems GEOSS. 10-Year Implementation Plan Reference Document.* https://www.earthobservations.org/documents/10-Year%20Implementation%20Plan.pdf

Gessner, U., J. Bliefernicht, M. Rahmann, and S. Dech. 2012. "Land Cover Maps for Regional Climate Modeling in West Africa – A Comparison of Datasets." Poster at the 32nd EARSeL Symposium, Mykonos, May 23–25.

Gong, P., J. Wang, L. Yu, Y. Zhao, Y. Zhao, L. Liang, Z. Niu, X. Huang, H. Fu, S. Liu, C. Li, X. Li, W. Fu, C. Liu, Y. Xu, X. Wang, Q. Cheng, L. Hu, W. Yao, H. Zhang, P. Zhu, Z. Zhao, H. Zhang, Y. Zheng, L. Ju, Y. Zhang, H. Chen, A. Yan, J. Guo, L. Yu, L. Wang, X. Liu, T. Shi, M. Zhu, Y. Chen, G. Yang, P. Tang, B. Xu, C. Giri, N. Clinton, Z. Zhu, J. Chen, and J. Chen. 2013. "Finer Resolution Observation and Monitoring of Global Land Cover: First Mapping Results with Landsat TM and ETM+ Data." *International Journal of Remote Sensing* 34 (7): 2607–2654. doi:10.1080/01431161.2012.748992.

Gupta, A. 2009. "Geology and Landforms of the Mekong Basin." In *The Mekong. Biophysical Environment of an International River Basin*, edited by I. C. Campbell, 29–52. New York: Elsevier.

Haines-Young, R. 2009. "Land Use and Biodiversity Relationships." *Land Use Policy* 26 (Suppl 1): S178–S186. doi:http://dx.doi.org/10.1016/j.landusepol.2009.08.009.

Hansen, M., R. DeFries, J. R. G. Townshend, and R. Sohlberg. 2000. "Global Land Cover Classification at 1 km Resolution Using a Decision Tree Classifier." *International Journal of Remote Sensing* 21: 1331–1365.

Hansen, M. C., and B. Reed. 2000. "A Comparison of the IGBP DISCover and University of Maryland 1 Km Global Land Cover Products." *International Journal of Remote Sensing* 21 (6–7): 1365–1373.

Herold, M., P. Mayaux, C. E. Woodcock, A. Baccini, and C. Schmullius. 2008. "Some Challenges in Global Land Cover Mapping: An Assessment of Agreement and Accuracy in Existing 1 Km Datasets." *Remote Sensing of Environment* 112 (5): 2538–2556. doi:10.1016/j.rse.2007.11.013.

Herold, M., C. E. Woodcock, A. Di Gregorio, P. Mayaux, A. S. Belward, and J. Latham. 2006. "A Joint Initiative for Harmonization and Validation of Land Cover Datasets." *IEEE Transactions on Geoscience and Remote Sensing* 44 (7): 1719–1727.

House-Peters, L. A., and H. Chang. 2011. "Modeling the Impact of Land Use and Climate Change on Neighborhood-Scale Evaporation and Nighttime Cooling: A Surface Energy Balance Approach." *Landscape and Urban Planning* 103 (2): 139–155. doi:http://dx.doi.org/10.1016/j.landurbplan.2011.07.005.

Hutyra, L. R., B. Yoon, J. Hepinstall-Cymerman, and M. Alberti. 2011. "Carbon Consequences of Land Cover Change and Expansion of Urban Lands: A Case Study in the Seattle Metropolitan Region." *Landscape and Urban Planning* 103 (1): 83–93. doi:http://dx.doi.org/10.1016/j.landurbplan.2011.06.004.

Jung, M., K. Henkel, M. Herold, and G. Churkina. 2006. "Exploiting Synergies of Global Land Cover Products for Carbon Cycle Modeling." *Remote Sensing of Environment* 101: 534–553.

Karnieli, A., U. Gilad, M. Ponzet, T. Svoray, R. Mirzadinov, and O. Fedorina. 2008. "Assessing Land-Cover Change and Degradation in the Central Asian Deserts Using Satellite Image Processing and Geostatistical Methods." *Journal of Arid Environments* 72 (11): 2093–2105. doi:http://dx.doi.org/10.1016/j.jaridenv.2008.07.009.

Koschke, L., C. Fürst, S. Frank, and F. Makeschin. 2012. "A Multi-Criteria Approach for an Integrated Land-Cover-Based Assessment of Ecosystem Services Provision to Support Landscape Planning." *Ecological Indicators* 21: 54–66. doi:http://dx.doi.org/10.1016/j.ecolind.2011.12.010.

Kuenzer, C. 2013. Field Note: Threatening Tonle Sap: Challenges for Southeast-Asia's Largest Freshwater Lake, 29–31. The Pacific Geographies (PG). ISSN 2196–1468.

Kuenzer, C., I. Campbell, M. Roch, P. Leinenkugel, T. Vo Quoc, and S. Dech. 2012. "Understanding the Impacts of Hydropower Developments in the Context of Upstream-Downstream Relations in the Mekong River Basin." *Sustainability Science*. doi:10.1007/s11625-012-0195-z.

Kuenzer, C., and S. Dech. 2013. "Theoretical Background of Thermal Infrared Remote Sensing." In *Thermal Infrared Remote Sensing – Sensors, Methods, Applications*, edited by C. Kuenzer, and S. Dech, 572 pp. Remote Sensing and Digital Image Processing Series, Volume 17. ISBN 978-94-007-6638-9, pp–1–26.

Kuenzer, C., H. Guo, P. Leinenkugel, J. Huth, X. Li, and S. Dech. 2013. "Flood Mapping and Flood Dynamics of the Mekong Delta: An ENVISAT-ASAR-WSM Based Time Series Analyses." *Remote Sensing* 5: 687–715. doi:10.3390/rs5020687.

Latifovic, R., Z.-L. Zhu, J. Cihlar, C. Giri, and I. Olthof. 2004. "Land Cover Mapping of North and Central America – Global Land Cover 2000." *Remote Sensing of Environment* 89: 116–127.

Leinenkugel, P., C. Kuenzer, N. Oppelt, and S. Dech. 2013. "Characterisation of Land Surface Phenology and Land Cover Based on Moderate Resolution Satellite Data in Cloud Prone Areas– A Novel Product for the Mekong Basin." *Remote Sensing of Environment* 136: 180–198. doi:10.1016/j.rse.2013.05.004.

Leinenkugel, P., M. L. Wolters, C. Kuenzer, N. Oppelt, and S. Dech. 2014. "Sensitivity Analysis for Predicting Continuous Fields of Tree-Cover and Fractional Land-Cover Distributions in Cloud-Prone Areas." *International Journal of Remote Sensing* 35 (8): 2797–2819. doi:10.1080/01431161.2014.890302.

Liu, Y., Q. Zhuang, M. Chen, Z. Pan, N. Tchebakova, A. Sokolov, D. Kicklighter, J. Melillo, A. Sirin, G. Zhou, Y. He, J. Chen, L. Bowlin, D. Miralles, and E. Parfenova. 2013. "Response of Evapotranspiration and Water Availability to Changing Climate and Land Cover on the Mongolian Plateau During the 21st Century." *Global and Planetary Change* 108: 85–99. doi:http://dx.doi.org/10.1016/j.gloplacha.2013.06.008.

Matthews, E. 1983. "Global Vegetation and Land Use: New High-Resolution Data Bases for Climate Studies." *Journal of Climate and Applied Meteorology* 22: 474–487.

Mayaux, P., H. Eva, J. Gallego, A. H. Strahler, M. Herold, S. Agrawal, S. Naumov, E. E. De Miranda, C. M. Di Bella, C. Ordoyne, Y. Kopin, and P. S. Roy. 2006. "Validation of the Global Land Cover 2000 Map." *IEEE Transactions on Geoscience and Remote Sensing* 44 (7): 1728–1739.

McCallum, I., M. Obersteiner, S. Nilsson, and A. Shvidenko. 2006. "A Spatial Comparison of Four Satellite Derived 1 Km Global Land Cover Datasets." *International Journal of Applied Earth Observation and Geoinformation* 8: 246–255.

Molina, A., G. Govers, V. Vanacker, J. Poesen, E. Zeelmaekers, and F. Cisneros. 2007. "Runoff Generation in a Degraded Andean Ecosystem: Interaction of Vegetation Cover and Land Use." *CATENA* 71 (2): 357–370. doi:http://dx.doi.org/10.1016/j.catena.2007.04.002

Nelson, R., and B. Holben. 1986. "Identifying Deforestation in Brazil Using Multiresolution Satellite Data." *International Journal of Remote Sensing* 7 (3): 429–448.

Neumann, K., M. Herold, A. Hartley, and C. Schmullius. 2007. "Comparative Assessment of CORINE2000 and GLC2000: Spatial Analysis of Land Cover Data for Europe." *International Journal of Applied Earth Observation and Geoinformation* 9: 425–437.

Peng, T., and S.-J. Wang. 2012. "Effects of Land Use, Land Cover and Rainfall Regimes on the Surface Runoff and Soil Loss on Karst Slopes in Southwest China." *CATENA* 90: 53–62. doi:http://dx.doi.org/10.1016/j.catena.2011.11.001.

Pérez-Hoyos, A., F. J. García-Haro, and J. San-Miguel-Ayanz. 2012a. "A Methodology to Generate a Synergetic Land-Cover Map by Fusion of Different Land-Cover Products." *International Journal of Applied Earth Observation and Geoinformation* 19: 72–87.

Pérez-Hoyos, A., F. J. García-Haro, and J. San-Miguel-Ayanz. 2012b. "Conventional and Fuzzy Comparisons of Large Scale Land Cover Products: Application to CORINE, GLC2000, MODIS and GlobCover in Europe." *ISPRS Journal of Photogrammetry and Remote Sensing* 74: 185–201.

Petersen, G. W., K. F. Connors, D. A. Miller, R. L. Day, and T. W. Gardner. 1987. "Aircraft and Satellite Remote Sensing of Desert Soils and Landscapes." *Remote Sensing of Environment* 23 (2): 253–271. doi:http://dx.doi.org/10.1016/0034-4257(87)90041-1.

Ran, Y., X. Li, and L. Lu. 2010. "Evaluation of Four Remote Sensing Based Land Cover Products Over China." *International Journal of Remote Sensing* 31: 391–401.

Renaud, F., and C. Kuenzer. 2012. *The Mekong Delta System*, 133–166. Interdisciplinary Analyses of a River Delta. Springer. ISBN: 978-94-007-3961-1.

Rittenhouse, C. D., and A. R. Rissman. 2012. "Forest Cover, Carbon Sequestration, and Wildlife Habitat: Policy Review and Modeling of Tradeoffs Among Land-Use Change Scenarios." *Environmental Science & Policy* 21: 94–105. doi:http://dx.doi.org/10.1016/j.envsci.2012.04.006.

Rojas, C., J. Pino, C. Basnou, and M. Vivanco. 2013. "Assessing Land-Use and -Cover Changes in Relation to Geographic Factors and Urban Planning in the Metropolitan Area of Concepción (Chile). Implications for Biodiversity Conservation." *Applied Geography* 39: 93–103. doi:http://dx.doi.org/10.1016/j.apgeog.2012.12.007.

Schilling, K. E., K.-S. Chan, H. Liu, and Y.-K. Zhang. 2010. "Quantifying the Effect of Land Use Land Cover Change on Increasing Discharge in the Upper Mississippi River." *Journal of Hydrology* 387 (3–4): 343–345. doi:http://dx.doi.org/10.1016/j.jhydrol.2010.04.019.

Schneider, A., M. A. Friedl, and D. Potere. 2010. "Mapping Global Urban Areas Using MODIS 500-M Data: New Methods and Datasets Based on 'Urban Ecoregions'." *Remote Sensing of Environment* 114 (8): 1733–1746. doi:10.1016/j.rse.2010.03.003.

Schneider, J., G. Grosse, and D. Wagner. 2009. "Land Cover Classification of Tundra Environments in the Arctic Lena Delta Based on Landsat 7 ETM+ Data and Its Application for Upscaling of Methane Emissions." *Remote Sensing of Environment* 113 (2): 380–391. doi:http://dx.doi.org/10.1016/j.rse.2008.10.013.

Sexton, J. O., X. Song, M. Feng, P. Noojipady, A. Anand, C. Huang, D. Kim, C. M. Collins, S. Channan, C. DiMiceli, and J. R. Townshend. 2013. "Global, 30-M Resolution Continuous Fields of Tree Cover: Landsat-Based Rescaling of MODIS Vegetation Continuous Fields with Lidar-Based Estimates of Error." *International Journal of Digital Earth*. doi:10.1080/17538947.2013.786146.

Shi, Z. H., L. Ai, X. Li, X. D. Huang, G. L. Wu, and W. Liao. 2013. "Partial Least-Squares Regression for Linking Land-Cover Patterns to Soil Erosion and Sediment Yield in Watersheds." *Journal of Hydrology* 498: 165–176. doi:10.1016/j.jhydrol.2013.06.031.

Sohl, T. L., B. M. Sleeter, Z. Zhu, K. L. Sayler, S. Bennett, M. Bouchard, R. Reker, T. Hawbaker, A. Wein, S. Liu, R. Kanengieter, and W. Acevedo. 2012. "A Land-Use and Land-Cover Modeling Strategy to Support a National Assessment of Carbon Stocks and Fluxes." *Applied Geography* 34: 111–124. doi:http://dx.doi.org/10.1016/j.apgeog.2011.10.019.

Tchuenté, A. T. K., J.-L. Roujean, and S. M. de Jong. 2011. "Comparison and Relative Quality Assessment of the GLC2000, GLOBCOVER, MODIS and ECOCLIMAP Land Cover Data Sets at the African Continental Scale." *International Journal of Applied Earth Observation and Geoinformation* 13 (2): 207–219.

Townshend, J. R., J. G. Masek, C. Huang, E. Vermote, F. Gao, S. Channan, J. O. Sexton, M. Feng, R. Narasimhan, D. Kim, K. Song, D. Song, X. Song, P. Noojipady, B. Tan, M. C. Hansen, M. Li, and R. E. Wolfe. 2012. "Global Characterization and Monitoring of Forest Cover Using Landsat Data: Opportunities and Challenges." *International Journal of Digital Earth* 5 (5): 373–397. doi:10.1080/17538947.2012.713190.

Tsegaye, D., S. R. Moe, P. Vedeld, and E. Aynekulu. 2010. "Land-Use/Cover Dynamics in Northern Afar Rangelands, Ethiopia." *Agriculture, Ecosystems & Environment* 139 (1–2): 174–180. doi:10.1016/j.agee.2010.07.017.

Verburg, P. H., K. Neumann, and L. Nol. 2011. "Challenges in Using Land Use and Land Cover Data for Global Studies." *Global Change Biology* 17: 974–989. doi:10.1111/j.1365–2486.2010.02307.x.

Vo, Q. T., C. Kuenzer, Q. M. Vo, F. Moder, and N. Oppelt. 2012. "Review of Valuation Methods for Mangrove Ecosystem Services." *Ecological Indicators* 23: 431–446. doi:http://dx.doi.org/10.1016/j.ecolind.2012.04.022.

Wang, H., X. Li, H. Long, Y. Qiao, and Y. Li. 2011. "Development and Application of a Simulation Model for Changes in Land-Use Patterns Under Drought Scenarios." *Computers & Geosciences* 37 (7): 831–843. doi:10.1016/j.cageo.2010.11.014.

Waser, L. T., and M. Schwarz. 2006. "Comparison of Large-Area Land Cover Products with National Forest Inventories and CORINE Land Cover in the European Alps." *International Journal of Applied Earth Observation and Geoinformation* 8: 196–207.

Wilson, M. F., and A. Henderson-Sellers. 1985. "A Global Archive of Land Cover and Soils Data for Use in General Circulation Climate Models." *Journal of Climatology* 5: 119–143.

Wu, W., R. Shibasaki, P. Yang, L. Ongaro, Q. Zhou, and H. Tang. 2008. "Validation and Comparison of 1 km Global Land Cover Products in China." *International Journal of Remote Sensing* 29 (13): 3769–3785. doi:10.1080/01431160701881897.

Yan, B., N. F. Fang, P. C. Zhang, and Z. H. Shi. 2013. "Impacts of Land Use Change on Watershed Streamflow and Sediment Yield: An Assessment Using Hydrologic Modelling and Partial Least Squares Regression." *Journal of Hydrology* 484: 26–37. doi:10.1016/j.jhydrol.2013.01.008.

Land-surface temperature dynamics in the Upper Mekong Basin derived from MODIS time series

C.M. Frey and C. Kuenzer

*German Remote Sensing Data Center (DFD), German Aerospace Center (DLR),
Oberpfaffenhofen, Germany*

Land surface temperature (LST) is an important indicator for climate variability and can be sensed remotely by satellites with a high temporal resolution on a broad spatial scale. In this research, Moderate Resolution Imaging Spectroradiometer (MODIS) LST is used to derive a 13 year time series on the Upper Mekong Basin (UMB), belonging to the People's Republic of China and the Republic of the Union of Myanmar, to analyse the spatial pattern and temporal development of LST. The data set shows the regular annual curve of surface temperatures with maximum values in summer and minimum values in winter. Average temperatures in the southern parts of the basin are higher than in the northern part. Spatial gradients between maximum and minimum LST as well as gradients between daytime and night-time LST are much lower in the southern parts than in the northern parts, which are characterized by a strong topography. The pixel-wise variability of monthly means was found to be in the range of ±4°C for most pixels in the daytime scenes, whereas the night-time scenes show a lower variability with most pixels in the range of ±1°C. The variability of LST in the northern areas clearly exceeds that in the southern areas. Some inter-annual variations occur, mainly during summer: in some years a two-peak distribution is found, which is explained by the generally low number of observations in the respective months. A primary challenge of optical satellite data in the UMB is cloud contamination in the summer months, where peak rainfall occurs. In the Mekong Highlands for instance, the average number of available daytime observations of MODIS LST in July is one observation per month only. It can be assumed that climate statistics calculated from such data is biased. In this context, two gap-filling algorithms were applied to two test areas for the year 2002 and results are discussed in the article. Another issue with MODIS LST data are day-to-day differences in the acquisition time. A temporal homogenization was applied to selected LST data, converting them to one fixed acquisition time. The converted data were compared to the original data set. No significant influence could be found.

1. Introduction

Land surface temperature (LST) is an important variable in the surface energy budget. The magnitude of the sensible heat flux and thus the air temperature is controlled by LST. This variable also determines the magnitude of the longwave radiation leaving the Earth's surface. Hence, LST is an important indicator for climate variability (Kuenzer and Dech 2013) and can be sensed remotely by satellites with a high temporal resolution on a broad spatial scale. As incoming shortwave radiation is the main governing factor of LST, it has a strong diurnal course and a distinct annual cycle. Furthermore, LST shows fluctuations in different time scales owing to changing cloud-cover and weather patterns. Besides this

34

temporal variability, LST has various spatial patterns, which are induced by different surface properties of the land-cover types and which partly complicate the measurements of LST with a remote sensor. The emissivity of the surface, the heat capacity of the surface soil layers, soil moisture, and thermal anisotropy produced by the structure of the surface elements are probably the most prominent factors that affect the spatial distribution of LST (Kuenzer and Dech 2013). The measurement of LST from space started in the 1980s (Landsat TM, AVHRR); currently, longer data sets are available that allow the application of time-series techniques (Jin and Dickinson 2010; Tian et al. 2012). Recent research tries to derive surface air temperature from LST data sets and climatologies (Shen and Leptoukh 2011; Benali et al. 2012). The LST time series can also be used, for instance, in ecology and epidemiology (Neteler 2010).

Owing to the large influence of incoming radiation on LST, a cloud-detection scheme must be added to any LST analysis. However, accurate cloud detection is still a main challenge for the community of optical satellite data users and cloud contamination is identified as a major problem in many applications (Brown et al. 2006; Fensholt et al. 2009; Frey, Kuenzer, and Dech 2012; Gutman and Masek 2012; Moreno-Ruiz et al. 2009; Leinenkugel, Kuenzer, and Dech 2013). Besides various attempts to improve reliable cloud detection in operational processing (e.g. Heidinger et al. 2012), the filling of cloud-masked pixels with numerical values is being researched. Poggio, Gimona, and Brown (2012) summarized the gap-filling methods into three categories: (1) combining observations from different satellites at different times; (2) spatial filtering, using information from neighbouring pixels in a moving window; and (3) temporal filtering, using the information of the same pixel in a proceeding cloud-free temporal window. A very simple example of temporal filtering is maximum value compositing (Holben 1986). For all three categories, a multitude of methods has been presented; however, all these methods were mainly developed for vegetation and snow variables and bands in the shortwave domain. LST strongly depends on solar radiation and changes with slope, exposition, and under cloud cover. Therefore, most of the described methods are not applicable for LST, although some methods were presented for LST. Jin (2000) describes a neighbouring-pixel approach for gap-filling LST data sets. In this approach, the LST of a cloudy pixel is related to the LST of a spatially or temporally neighbouring clear pixel through the surface energy balance equation. The approach further includes an adjustment derived from surface air temperature estimates and relies on parameterized diurnal cycle shapes and some ancillary information (soil moisture, vegetation, cloud properties, season, and latitude). Neteler (2010) presents a simpler algorithm that reconstructs LST based on temperature gradient maps. Recently, Xu, Shen, and Wu (2013) presented the removal of cloud-cover effects by fitting the LST course with the 'Harmonic Analysis of Time Series' (HANTS) algorithm. Both algorithms from Neteler (2010) and Xu, Shen, and Wu (2013) provide time series that can be understood as clear-sky LST.

In this research, a 13 year Moderate Resolution Imaging Spectroradiometer (MODIS) LST time series is used on the Upper Mekong Basin (UMB) to analyse the temporal and spatial dynamics of LST. The cloud information of the MODIS mask delivered within the product was used. Recent research (Westermann, Langer, and Boike 2011; Leinenkugel, Kuenzer, and Dech 2013) has shown that the MODIS cloud mask may be incomplete in some cases, leaving cirrus or cloud-contaminated pixels unflagged. Contaminated pixels usually inhibit lower LST values than clear pixels (Hachem, Duguay, and Allard 2012). As a first assumption, it is anticipated that such erroneous values do not affect long-term analyses significantly; however, they may strongly affect the results of short-term analyses. Analysis will therefore include the possibility of undetected clouds.

To evaluate the influence of cloud gaps on the results of the time-series investigation, two gap-filling approaches were applied to two test areas and the results were analysed. First, a gradient-based method comparable to Neteler (2010) is used. Second, gaps were filled using estimates from ECMWF era-interim analysis data. Both methods are explained in Section 4.

2. Area of interest

The area of interest is the UMB. The major part of the UMB belongs to the People's Republic of China; only the southernmost tip belongs to the Republic of the Union of Myanmar (Figure 1). According to Leinenkugel et al. (2013), the UMB consists of three

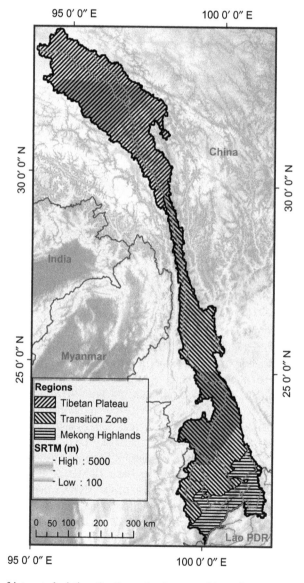

Figure 1. Area of interest depicting the three physiogeographic regions of the UMB: the Tibetan Plateau, the transition zone, and the Mekong Highlands. Also the extent of two test areas is given in grey.

physiographic homogeneous regions: (1) the Tibetan Plateau, (2) a transition zone (in the article called the Lancangjiang Basin), and (3) parts of the Mekong Highlands. The Tibetan Plateau is a vast, elevated plateau with high-altitude, polar tundra climate, with alpine grasslands above the treeline, and needle-leaved forests and shrubland below it. Here, the Mekong River has its origin at 5200 m. The transition zone shows a strong elevation gradient with numerous high mountains (3500–6740 m.a.s.l. – He et al. 2009) and deep valleys, mostly covered by coniferous forests and shrublands. Below 2000–2500 m, the forests change to broadleaved evergreen vegetation. Also larger portions of cropland are found here. The Mekong Highlands are characterized by high proportions of evergreen broadleaved forests, interspersed by small-scale fields from cropping and shifting cultivation practices (Leinenkugel et al. 2013). According to Costa-Cabral et al. (2008), who used the USDA soil texture classes, soils in the UMB mostly consist of loam and sandy clay loam. In the Tibetan Plateau also sandy loam, clay, and clay loam occur. Owing to the mountainous environment with steep slopes and unstable geology, soils undergo strong erosion (He et al. 2009). The upper Mekong (in China called Lancang) is a base for hydropower and mineral resource development, but is also key to ecological conservation. A series of development programmes, e.g. the China Great Western Development Programme or the Greater Mekong Subregional Economic Cooperation Programme, has led to dramatic changes in the course of the river (He et al. 2009; Kuenzer et al. 2012).

In this research, only the parts inside the UMB (black outline in Figure 1) are taken into account. Pixels lying outside the UMB were not considered in any calculations. The grey areas in Figure 1 depict the two test areas that were used for the gap-filling applications. The first test area belongs to the transition zone and the second to the Tibetan Plateau. The test areas were selected to represent two typical regions of the UMB. Their shapes arise from clipping the UMB from respective rectangular tiles, in which the data is archived locally.

3. Data

Input for this analysis is the gridded MODIS LST data from the Terra satellite from version 5 (MOD1A1v5) from 2000 to 2012. The product is delivered daily (one daytime and one night-time scene) with 1 km resolution in sinusoidal projection. The generalized split-window algorithm is used to derive LST from bands 31 and 32 and known emissivities (Wan and Dozier 1996). The accuracy of the MODIS LST product is 1 K, although higher errors may occur at larger viewing angles, and in semi-arid and arid regions due to the effects of heavy dust aerosols and difficulties in emissivity estimation (MODIS Land Team Validation, http://landval.gsfc.nasa.gov/). Besides an additional post-seasonal cloud screening (Neteler 2010), which is described in Section 4, the MODIS data are used as they are, assuming that possible effects resulting from thermal anisotropy and uncertainties in emissivity estimation have only minor influence on the time-series analysis. Furthermore, long-term monthly mean enhanced vegetation index (EVI) values were used. These data were derived from an 11 year time series of the MODIS product MCD09A1 in another research work (Leinenkugel et al. 2013).

To analyse the spatial distribution of LST, a digital elevation model (DEM) from the Shuttle Radar Topography Mission (SRTM) is used. The DEM was projected into the sinusoidal projection of the MODIS data and re-sampled to 1 km resolution.

Furthermore, a set of skin temperature data was used for a gap-filling method as described in Section 5. The skin temperature was taken from the Interim Re-analysis Surface Analysis Data Set of the European Centre for Medium-Range Weather Forecasts (ECMWF). The skin temperature is defined as the temperature of a surface (interface between soil and atmosphere) at radiative equilibrium and is therefore equivalent to remotely sensed LST.

4. Methodology

In the MODIS LST product, clouded pixels are already masked out. However, in some instances, e.g. at the edges of clouds or over bare surfaces, there might be pixels with a partial cloud cover that was not detected. To minimize the number of such pixels and to ensure that only good-quality pixels are used in the statistics, only LST pixels that were flagged as 'good quality' in the accompanying quality layer were used for this research. In addition to this quality procedure, a post-processing scheme proposed by Neteler (2010) was implemented. This scheme uses a 'lower boundary' as the threshold for the screening of valid pixels. The lower boundary is calculated separately for each month based on quartiles, which are derived pixel-wise from all valid values of a month.

To estimate the direct influence of missing data on the results, two methods were developed, whose output can be used to fill the gaps. In the following, the two methods are shortly described.

4.1 *Gradient-based method*

In the first method, the gradient-based method – which is similar to the method applied by Neteler (2010) – each pixel is compared to all other pixels of the input area.

Thereby, series of gradients between the considered pixel and all other pixels in the subset are calculated for each day of a month, resulting in an array of maximal 31 gradients in case of a month with 31 days for each pixel. From these arrays, a selection of 'good pixels' with most stable gradients in each respective month is conducted on the basis of statistical values as follows. For each series of gradients (between the considered pixel and another pixel in the subset), the mean and the standard deviation of all values inside the 25- and 75-percentile are calculated and saved to a file. In a second step, for each pixel in the image, surrounding pixels with stable gradients (to the selected pixel) are searched using the information from the mean and the standard deviation saved to a file in the first step. The selected suitable pixels are then used for LST construction. As long as five or more gradients are available for the reconstruction of an LST value, all its gradients whose standard deviation is lower than 0.1 times the mean are marked as suitable. In case fewer gradients are available, then the gradients whose standard deviation is lower than 0.3 times the mean are marked as suitable. Finally, new values for each day are calculated using an average of the suitable gradients and their actual LST counterpart. The gradient based method gets along with only the MODIS LST data and does not need any auxiliary data. However, the method uses excessive computing power and is not yet appropriate for operational use. The gradient-based method creates a set of clear-sky LSTs.

4.2. *Era-interim-based method*

The second method uses ECMWF Era-Interim analysis skin temperature data to estimate the missing values. The data have been obtained from the ECMWF data server. The era-interim data comes in 0.125° spatial resolution and in four time steps per day (00:00, 06:00, 12:00, and 18:00 UTC). The data is interpolated to daily curves with 6 min time steps using the cubic spline interpolation method. For each pixel the mean difference of one month between the valid MODIS LST pixels and the interpolated era-interim temperatures is calculated. This difference is then added to the actual interpolated era-interim value, which is then used to estimate the value of a cloudy pixel. The resulting data set is not a clear-sky data set, as the era-interim temperatures do include cloudy grid points. So, in cases of extended and permanent cloudiness the input values represent LST under clouds, although the monthly difference input parameter might represent a mixture between clear sky and cloudy sky.

All calculations were performed in the sinusoidal projection of MODIS.

5. Results

5.1. *Spatial pattern distribution of LST*

The distribution of LST in the UMB is highly structured and exhibits large gradients. In the following, mean LST values over the 13 year period are provided separately for the three regions presented in Section 2. The Tibetan Plateau shows very cool surface temperatures. Minimum mean daytime LST is found in February with −16°C. A night-time minimum is found in January with −25°C. Nevertheless, in some valleys of the region next to the transition zone, extremely high LSTs occur in summer. Maximum mean daytime LST in June is found to be 48°C. In the night, maximum mean LST of July is 20°C. Also in the transition zone extremely low surface temperatures are found along with strong LST gradients present in the steep valleys. At the bottom of the valley high temperatures may occur. The magnitude of minimum and maximum mean LST is similar to that of the Tibetan Plateau, although minima are a few degrees Celsius higher and maxima a few degrees Celsius lower. The Mekong Highlands generally feature higher LST than the other two regions. Maximum mean daytime LST is only 34°C in June; however, minimum mean daytime LST is 13°C in January and December. The median of monthly mean LST of the Mekong Highlands is always higher than the one of the Tibetan Plateau and mostly higher than the median of monthly mean LST of the transition zone.

The mean spatial gradient between maximum and minimum monthly means for the Mekong Highlands is 3°C in the daytime scenes, whereas in the night-time scenes it is 2°C. In the transition zone and the Tibetan Plateau, the daytime spatial gradient is 21°C resp. 23°C and the night-time gradient is 16°C resp. 9°C.

The topography appears in the pattern of the LST distribution, influencing it through height and exposition. This influence is found in all months and for the daytime and night-time scenes. The topography pattern induced by height is stronger in the night-time scenes than in the daytime scenes, where LST is also strongly influenced by exposition. Very large gradients in the LST distribution are predominantly found in the summer months. In the Tibetan Plateau in May and June, for instance, the daytime gradient between minimum and maximum mean LST is 47°C. Winter scenes appear more homogeneous, although gradients are still large: the same gradient in December is 32°C. The Mekong Highlands do not show such extreme gradients. They range from 12°C to 19°C only during daytime. Figure 2 shows the map of the mean daytime and night-time LST distribution over the whole 13 year period for the months of June and December.

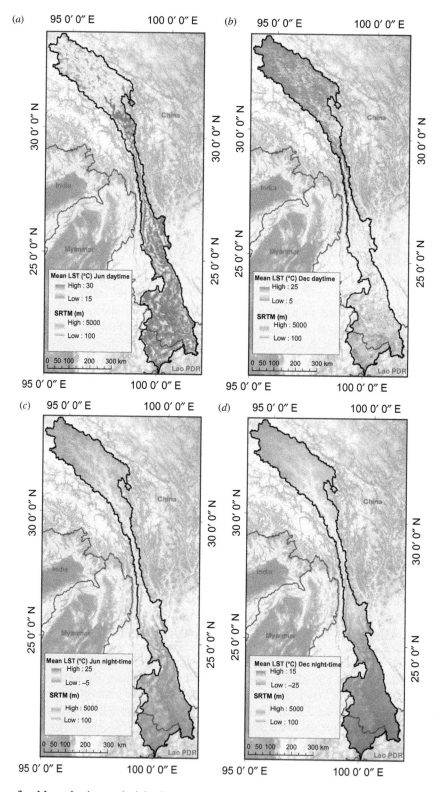

Figure 2. Mean daytime and night-time LST pattern distribution for the years 2000–2012. The months June and December are given.

5.2. *Time-series analysis*

The surface-temperature regime in the UMB is governed by the incoming solar radiation, which manifests in a distinct annual course with peak temperatures around June. In some years however, mainly in the daytime scenes, a two-peak distribution is found with a small trough during the summer months. This pattern is found in all regions and in the daytime and night-time means (Figure 3). The annual course is different for various parts of the UMB. As was shown before, in the northern parts – the Tibetan Plateau – LST shows the lowest temperatures; however, there was a large difference in monthly regional means (20–22°C, daytime and night-time) between summer and winter values. In the southern parts (transition zone and Mekong Highlands), LST is generally higher, with a slightly smoother annual curve. This effect is stronger for the night-time scenes than for the daytime scenes. Here differences between summer and winter monthly regional means reach 16°C in the daytime scenes and 10–11°C in the night-time scenes. Annual curves get continuously smoother towards the south, which is in line with the fact that seasonal differences become weaker when getting closer towards the equator. Figure 3 shows the mean monthly daytime and night-time LST separated for each region (Figure 1) for the whole time period.

Although the annual courses of LST are rather continuous, there are areas with months that strongly deviate from the long-time monthly mean. In the following, the difference between actual monthly means and long-term monthly means will be addressed as

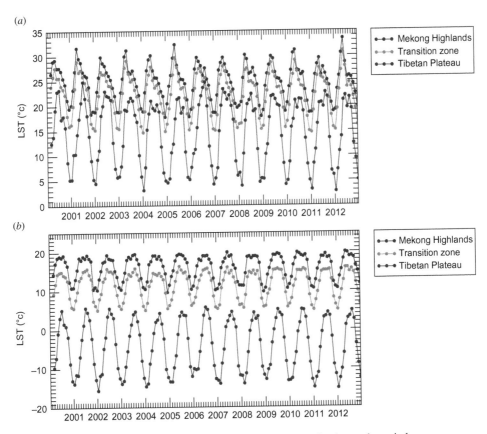

Figure 3.   Regional mean daytime (*a*) and night-time (*b*) LST for the study period.

deviations. In the climatological community, also the term 'anomaly' is used for this difference (NOAA 2013). Large monthly deviations are found in the Tibetan Plateau (regional monthly mean deviation up to 5°C, maximum monthly mean deviation per pixel: 24°C) and of the transition zone (regional monthly mean deviation up to 4°C, maximum monthly mean deviation per pixel: 17°C) in the daytime scenes. The monthly mean deviations of the Mekong Highlands region reach 3°C with a maximum monthly mean deviation of 13°C per pixel. Night-time deviations are generally lower, regional monthly mean deviations are equal to or below 4°C in all regions. However, monthly mean deviations per pixel may be much higher (transition zone: 31°C), which is due to cloud-contaminated pixels that were wrongly labelled as 'Good quality' and were also not detected by the post-processing scheme. Figure 4 shows the monthly mean deviations per region.

Figure 4 shows that the deviations are not equal for the three regions. Also within the regions, there are large differences in the long-term variability. Figures 5 and 6 shows the variability of the whole UMB separated for daytime and night-time scenes and each month. Hereby, variability is defined as the standard deviation from the mean monthly deviation (2000–2012) per pixel. Although the variability in the night-time scenes is rather small (mostly below 1°C), in the daytime scenes a decoupling of the Tibetan

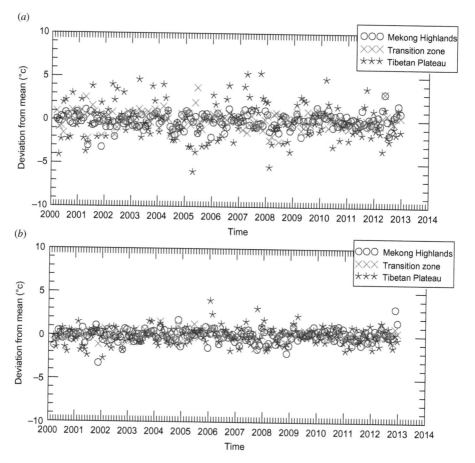

Figure 4.   Mean daytime (*a*) and night-time (*b*) deviations from regional mean LST for the study period.

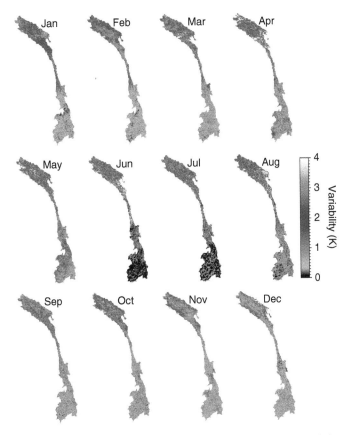

Figure 5.   Daytime monthly variability of LST for the years 2000–2012. North is up.

Plateau is again clearly visible, where high variability is found (mostly between 1°C and 4°C). Extreme variability is mainly found in the peripheral areas of the Tibetan Plateau. The Mekong Highlands show a rather low (mostly below 1°C) variability in LST deviations. The magnitude of the variability follows the magnitude of the deviations: in areas with high deviations, high variability also occurs.

6.   Discussion of factors influencing the spatial and temporal LST distribution

6.1.   Time-series results

The distribution of LST in the UMB shows a distinct spatial pattern, which is considerably influenced by the topography of the area. Correlation coefficients r^2 between the long-term 13 year monthly LST means and the topographic height, which was derived from the SRTM data set, reveal a strong relation, especially for the night-time monthly scenes. r^2 of the night-time scenes ranges from 0.90 to 0.94. The r^2 of the daytime monthly scenes are lower and range from 0.40 to 0. 77. The correlation of the daytime scenes is stronger in the winter month; low correlations around 0.5 are solely found in the summer months. The high correlation during night can only be reached as solar effects from topography induced by slope and exposition are minimal during night. During daytime, especially in summer, when the Sun altitude of the Sun is close to the zenith, these effects are maximal and superimpose on the height effect.

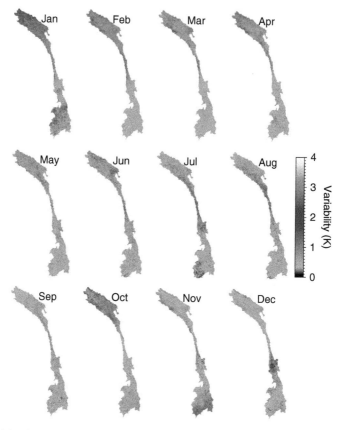

Figure 6. Night-time monthly variability of LST for the years 2000–2012. North is up.

Another factor that generally explains LST distribution is the latitude. The UMB stretches from almost 20° N to almost 34° N. The solar irradiation gets stronger from the north to south, which is accompanied by higher LST values towards the south. This inter-regional effect, however, is covered by the strong topography effect and not visible in the data. Additional local-scale effects that may influence the spatial LST distribution are the luv and lee effects of the local wind regime as well as snow cover, snow melt, and surface cover (bare soils vs. vegetated areas). The last and mostly local relation can be investigated using a vegetation index, comparing it with the LST pattern distribution. Here, the long-term monthly mean of EVI in correlation to the long-term monthly mean of LST is used. In the night-time scenes, a correlation coefficient r^2 of 0.7 was found for all months. In the daytime scenes, r^2 ranges from 0.2 to 0.6 with minimum values found in summer. When correlating the long-term mean EVI with the long-term mean LST for discrete height ranges, r^2 decreases strongly, suggesting that the above-found correlation is mainly due to the height dependence of both variables. It was also analysed whether there is a temporal relation between LST deviations and EVI deviations. The EVI deviations are thereby defined as the difference between actual means and long-term means, calculation is analogue to the LST deviations. It was found that in wintertime positive daytime LST deviations (=warmer temperatures) are weakly correlated with positive EVI deviations (=denser vegetation), whereas negative LST deviations (=cooler temperatures) occur simultaneously with negative EVI deviations (less-dense vegetation), which could be explained by a stronger vegetation growth in warmer years ($r^2 < 0.4$ in daytime scenes). In the night-time data however, the correlation was very weak ($r^2 < 0.2$). It must be noted that the daytime and night-time LST deviations are correlated only to a certain extent (r^2 ranges from 0.1 to 0.7), which may partly be explained by the quality of input LST data (undetected clouds, known limitations of the MODIS LST algorithm).

Apart from year-to-year variations, evidence for trends spanning the 13 year period was searched for in both daytime and night-time monthly mean data by regressing the LST time series with time. However, r^2 of most pixels was low (r^2 ranges from 0.0 to 0.9, mean r^2 is 0.1). A 13 year time series is not sufficient to deduct trends in a sound statistical manner. As such, the findings of a warming of the Tibetan Plateau (You et al. 2010; Salama et al. 2012) cannot be confirmed with this data set.

6.2. *Influence of cloud coverage on the results*

Optical satellite images are unfortunately often cloud-contaminated. Strong contamination hinders the spatial and temporal analysis of LST. To evaluate the influence of cloud gaps on the above-mentioned results, cloud coverage of the UMB was analysed and the two gap-filling methods, which are described in Section 4, were applied.

Cloud coverage in the UMB shows a distinct annual pattern with strong contamination in the summer month, where peak rainfall occurs. This pattern is enhanced in the southern region, where the influence of the summer monsoon is the strongest and therefore cloud contamination is the highest. Figure 7 shows the mean number of valid pixels for each of the three regions of the study area.

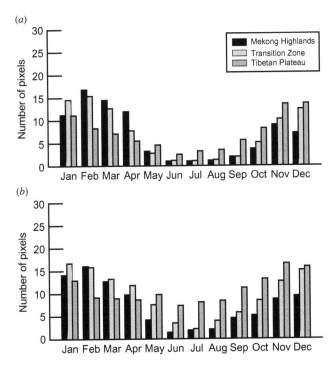

Figure 7.   Number of valid pixels per month in the (*a*) daytime and (*b*) night-time scenes. Monthly means averaged over the study period and the three regions are given.

The season of the two-peak distribution mentioned in Section 5.2 is summer, where very low numbers of valid observations occur. In some cases, these few pixels also show very low LST owing to the remaining cloud cover and therefore decrease the monthly means, leading to the aforementioned two-peak distribution. The negative deviation can range from a few degrees until more than 20°C below average temperatures.

The availability of cloud-free pixels is distributed heterogeneously across the UMB. The Tibetan Plateau has fewer available pixels than the Mekong Highlands from January to April; from May to December, the Mekong Highlands show stronger cloud cover than the Tibetan Plateau. The lowest availability of cloud-free pixels is in the Mekong Highlands from June to August, where the average number of available daytime measurements (per pixel) is one only. Generally, the night-time scenes have a slightly higher availability of cloud-free measurements; however, the increase in the average number of available measurements (per pixel) ranges between −2 and 6.

Leinenkugel, Kuenzer, and Dech (2013) have shown that the MODIS cloud mask (MOD35) is sometimes too conservative for Southeast Asia. Furthermore, they found that the cloud-cover statistics reveals irregular patterns following the land-cover distribution. These patterns arise from the cloud-retrieval scheme of MODIS. Leinenkugel, Kuenzer, and Dech (2013) propose using the MOD09 internal cloud mask, which seems to offer the better choice in this region. They further present a method to combine the MOD35 and MOD09 cloud masks, to obtain more clear-sky observations. However, the MODIS LST product (MOD1A1v5) only delivers clear-sky observations masked by the MOD35 product; thus, it is not possible to implement the recommendations of before-mentioned authors.

In the following, time-series statistics of two test areas, in which a gap-filling procedure using the two methods described in Section 4 was applied, is shown on the example of the year 2002. Prior to presenting the results, the accuracy of the two methods shall be evaluated. For this, the two gap-filling methods were also applied additionally to cloud-free pixels, which served as reference values. Hence, the modelled values could be compared to the cloud-free pixels on a pixel basis. In the case of the gradient-based method, all valid pixels in the tiles were compared to modelled ones. In the case of the era-interim-based method, a set of reference time points (20% of the valid pixels) was selected for the validation. These 20% were then gap filled with the information from the remaining 80% of valid pixels in the monthly time series. Table 1 shows the accuracy of the gradient-based method; Table 2 shows the accuracy of the era-interim-based method. The gradient-based method shows a high accuracy (RMS (root mean square) < 1.5 K, MAE (mean absolute error) < 0.8 K) in the first test area. In the second, the accuracy is a bit lower, with RMS < 2.9 K and MAE < 1.6 K. Generally, the gradient-based method underestimates the LST values slightly. The era-interim-based method has lower accuracies in the daytime scenes (RMS < 4.7 K, MAE < 3.7 K, both test areas), but better accuracy in the night-time scenes (RMS < 0.9 K, MAE < 0.4 K, both test areas). The accuracy discrepancy between daytime and night-time images may arise from the high

Table 1. Accuracy of the gradient-based method.

	First test area		Second test area	
	Daytime	Night-time	Daytime	Night-time
MAE (K)	0.83	0.53	1.62	1.00
RMS (K)	1.46	1.11	2.87	2.21
r^2	0.93	0.93	0.93	0.94

Table 2. Accuracy of the era-interim-based method.

	First test area		Second test area	
	Daytime	Night-time	Daytime	Night-time
MAE (K)	2.58	0.29	3.68	0.37
RMS (K)	3.17	0.55	4.71	0.86
r^2	0.63	0.98	0.79	0.99

spatial variability of LST during day in contrast to more even night-time distributions and to the low spatial resolution of the era-interim data set. MAE is the mean absolute error and calculated as the mean of all absolute differences between the original and the modelled LST values. The era-interim method showed a tendency to overestimate LST.

Both methods enhance strongly the availability of pixels. After filling up the gaps with the gradient-based method, about 2.5 times more pixels than in the original data set were available. The minimum number of available pixels per month is 11; the mean number of available pixels per month is 19. The era-interim-based method was very successful in the second test area. Almost all pixels were filled and the average number of available pixels per region and month is always the number of days in a month. In the first test area, however, only about 1.5 times more pixels than in the original data set were available. The minimum number of available pixels per month is 5; the mean number of available pixels per month is 11. The Era-Interim-based method depends on available measurements within one month and fails if not enough measurements are available. As the first test area has more cloudy days than the second, the Era-Interim-based method fails in this case to produce continuous time series.

To analyse the impact of gaps on the time-series statistics, daytime and night-time daily and monthly regional mean LST for the original MODIS LST data, and the two gap-filled data sets are compared. Generally, both gap-filled data sets follow the regular annual curve. However, it is noticeable that in the daily regional means the extremely low summer LST values from the original data set are not reproduced by the Era-Interim gap-filled data set. These low values are outliers that often have their origin in pixels with undetected clouds. By filling up the cloud gaps with 'good' values, the undetected cloudy pixels do not carry as much weight and hence the regional mean is higher. In gradient-based gap filling, however, similar low outliers are produced through error propagation. The gradient-based gap-filled data set showed especially in the second test area a slight underestimation of daily regional mean LST.

Moreover, the monthly regional mean LST of gradient-based gap filling is significantly lower than the monthly regional mean from the original data (RMS ranges from 2.0°C to 4.7°C, see Table 3). Generally, filled-in values from the gradient-based method are lower than 'normal, clear-sky' pixels, as they are retrieved from pixels in the neighbourhood of clouds on cloudy days, which themselves are lower due to the missing insolation and subsequent heating of the ground. As such, modelled LST values are expected to be lower than an average clear-sky LST. The modelled monthly means of the summer month contain more modelled values than the original values and are therefore lower than the monthly means from the original data. The existence of undetected cloudy pixels enhances this effect considerably. The era-interim-based gap-filled data set did not show such an underestimation in the summer month (see Figure 8). Table 4 shows the MAE and RMS values of the original and the modelled LST using the era-interim-based method.

To quantitatively evaluate the difference between the original and the gap-filled data sets, the RMS and the MAE of daily, regional means for the year 2002 were calculated for

Table 3. Difference between the original regional means and the regional means of the modelled LST using the gradient-based method.

	First test area		Second test area	
	Daytime	Night-time	Daytime	Night-time
MAE (°C) (daily data)	0.91	1.17	1.49	0.99
RMS (°C) (daily data)	1.52	2.49	2.27	1.61
MAE (°C) (monthly data)	2.68	2.11	4.15	1.45
RMS (°C) (monthly data)	3.33	4.48	4.72	1.95

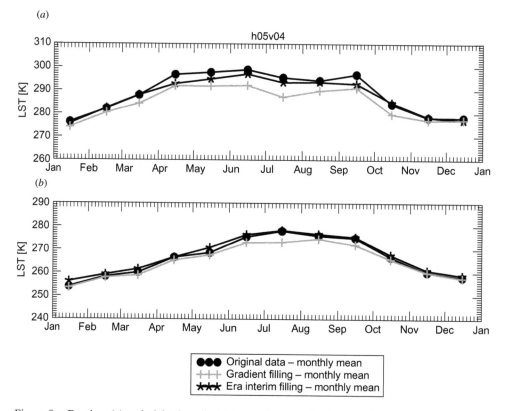

Figure 8. Daytime (*a*) and night-time (*b*) LST annual course for the second test area in 2002.

Table 4. Difference between the original regional means and the regional means of the modelled LST using the era-interim-based method.

	First test area		Second test area	
	Daytime	Night-time	Daytime	Night-time
MAE (°C) (daily data)	1.18	1.52	3.17	2.29
RMS (°C) (daily data)	1.92	3.41	4.79	3.65
MAE (°C) (monthly data)	0.30	0.08	1.52	0.99
RMS (°C) (monthly data)	0.54	0.11	2.03	1.19

the differences. Note that higher RMS/MAE indicates a higher error through cloud gaps in the original data set, assuming that the gap-filling values are true. The values suggest that the era-interim data set departs more strongly from the original data set than the gradient-based gap-filled data set.

Daily regional RMS of the original data set *versus* the era-interim data is 1.9°C and 4.8°C for the daytime scenes. For the night-time scenes RMS is 3.4°C and 3.7°C. These high values are in line with the finding that extreme outliers of the original data set are not reproduced in the era-interim gap-filled data set (Table 3). RMS of the original data set *versus* the gradient-based data ranges from 1.5°C to 2.5°C for the daytime and night-time scenes (Table 3) comparing the daily regional means.

The difference between the monthly means of the original and the gap-filled data sets is higher in the case of gradient-based gap filling and lower in the case of era-interim gap filling. RMS of the monthly regional means of the original data set *versus* the era-interim-based data ranges from 0.1°C to 2.0°C (both daytime and night-time scenes). RMS of the monthly regional means of the original data set *versus* the gradient-based data ranges from 2.0°C to 4.7°C.

According to Table 1, the gradient-based method works better in the first test area. Table 2 shows that the era-interim-based method has a very high accuracy in the night-time scenes. According to the gradient-based method, the influence of gaps is larger on the monthly regional means. However, results of the era-interim-based method, especially of the first test area, suggest that the influence on monthly means is lower. The large RMS of the gradient-based method mainly arises from the summer months, in which gradient-based gap filling is strongly influenced by misclassified pixels and enhances the mentioned two-peak distribution. It is therefore anticipated that the era-interim method produces more reliable annual courses as it eliminates negative outliers. As such it can be assumed that the influence of cloud gaps on long-term statistical values is not very severe and only partially affects the before-presented LST time series of the UMB. RMS of the regional means can be expected to be around 0.5–1°C (annual mean). However, major deviations are expected in summer. Higher deviations occur also on a daily basis (annual mean RMS of regional means around 3°C) in the summer month.

6.3. *Influence of time shift on the results*

MODIS LST data have day-to-day differences in the local acquisition time due to its wide field of view of approximately ±55° off nadir. The local (solar) time is a measure of the Sun's position relative to a locality and can be deducted from UTC acquisition time and longitude. Thereby one degree equals 4 min. For a given scene in the northern hemisphere, local time increases from west to east. The MODIS acquisition schedule has a 16 day repeat cycle, leading to different local times not only inside a scene but also inside the repeat day cycle. As LST shows a strong diurnal course, these differences may affect LST analyses, which depend on a temporally consistent and spatially comprehensive data set (Duan et al. 2011).

To assess the influence of this effect for this study, all cloud-free pixels were converted to a fixed local acquisition time using the methodology of the era-interim-based method. However, instead of filling existing gaps, only the actual pixel was substituted with the modelled value for the time of over-flight. These modelled values were then compared with the original LST values. The fixed times are 10:48 for the day scenes and 22:00 for the night scenes. These times are deducted from the average of all local times found in the time series, and were fitted to the temporal resolution of the interpolated era-interim data, which is 6 min.

The original and the corrected data sets are very well correlated and the effect on the daily mean is found to be small, especially during night. Statistical measures are provided

Table 5. Statistical measures of the comparison of the original LST data set and the time-adjusted LST data set to assess the influence of time shift.

	Daytime	Night-time
Mean (LST_{orig}) – Mean (LST_{norm})	$289.6 - 289.8 = 0.2$	$275.5 - 275.9 = -0.4$
RMS (K)	1.1	0.8
r^2	0.97	0.99

in Table 5. The differences between the mean of all original LST values and the mean of all normalized LST values are lower than 0.4 K for both daytime and night-time scenes. Considering the single differences between the two time series, there may be considerable differences. Maximum difference is 4.7 K in the daytime scene and 4.9 K in the night-time scenes. However, 80% of all differences of the daytime scenes are lower than 1.3 K and 90% are lower than 1.7 K. In the night-time scenes 80% are lower than 0.8 K and 90% are lower than 1.5 K. Some of these high differences are due to undiscovered clouds, as these calculations were performed without the post-processing scheme proposed by Neteler (2010). This means that the majority of the LST differences due to the time shift are lower than the claimed accuracy of the LST values itself.

7. Conclusions

Analysis of the LST time series has shown that the UMB shows two different climatic regimes concerning LST. The LST evolution of the Tibetan Plateau is characterized by strong annual cycles coupled with large gradients in inter-regional mean LST and inter-annual regional mean LST. The Mekong Highlands in contrast show much smoother annual courses and lower gradients of interregional and inter-annual LST. The transition zone finally is placed in between these two extremes. A description of these three areas can be found in Section 2. The topography was identified as the main factor explaining the LST pattern distribution in this region. The height is thereby the major factor, explaining 90% of the spatial variations of the night-time mean LST. In the daytime scenes other factors such as slope and explosion also influence the distribution.

Time-series analysis has shown that the spatial and temporal variability is different for the three regions. Highest temporal variability (=standard deviation of the monthly deviation from the long-term mean) is found on the Tibetan Plateau in the daytime monthly means. Extreme spatial variability is found in the peripheral areas of the Tibetan Plateau of the UMB. The Mekong Highlands region, on the contrary, shows low variability in the study period.

The UMB was also analysed in terms of possible evidence for trends in LST evolution. The analysis however showed that a 13 year time series is not sufficient to derive a proper statistics. The calculated values vary for each month and no consistent trend could be derived. Longer time series and more research are needed to clarify the LST evolution in this area.

The analysis of LST in the UMB is strongly hindered by the presence of clouds. Especially in the summer months, only a few valid pixels per month are available. To assess the influence of cloud gaps on monthly means, two cloud-gap-filling methods were applied. It was shown that the era-interim-based method was able to eliminate negative outliers in the time series, which originate in pixels with non-detected clouds. The gradient-based method was not able to perform this; in the data set gap-filled with the LST values estimated with the gradient-based method, negative outliers also appeared in the monthly regional means.

Overall influence of gaps on statistical values is large for daily regional means, but lower for monthly means. In the latter case the error magnitude is comparable to the accuracy of the daily product itself. However, during the summer months, monthly regional means might be biased considerably and exceed the stated accuracy of the daily product.

Another factor that could influence the statistics of time series is differences in local time of the MODIS acquisitions. This study showed that the differences cancel each other out when averaging over longer time periods. Considering a certain error of the correction method for the time differences, it is suggested that the differences in local time only marginally influence long time series.

Acknowledgements

The authors thank P. Leinenkugel for providing the physiographic regions of the Mekong Basin as well as the monthly EVI data to be used in this research.

Funding

The authors thank the German Ministry of Science and Education, BMBF, for funding the German–Vietnamese WISDOM project (www.wisdom.eoc.dlr.de).

References

Benali, A., A. C. Carvalho, J. P. Nunes, N. Carvalhais, and A. Santos. 2012. "Estimating Air Surface Temperature in Portugal Using MODIS LST Data." *Remote Sensing of Environment* 124: 108–121. doi:10.1016/j.rse.2012.04.024.

Brown, M. E., J. E. Pinz, K. Didan, J. T. Morisette, and C. J. Tucker. 2006. "Evaluation of the Consistency of Long-Term NDVI Time Series Derived from AVHRR, SPOT-Vegetation, SeaWiFS, MODIS, and Landsat ETM+ Sensors." *IEEE Transactions on Geoscience and Remote Sensing* 44 (7): 1787–1793. doi:10.1109/TGRS.2005.860205.

Costa-Cabral, M. C., J. E. Richey, G. Goteti, D. P. Lettenmaier, C. Feldkötter, and A. Snidvongs. 2008. "Landscape Structure and use, Climate, and Water Movement in the Mekong River Basin." *Hydrological Processes* 22: 1731–1746. doi:10.1002/hyp.6740.

Duan, S. B., H. Wu, N. Wang, X. M. Zhou, B. H. Tang, and Z. L. Li. 2011. "Preliminary Results of Temporal Normalization of MODIS Land Surface Temperature." Paper presented at the International Geoscience and Remote Sensing Symposium (IGARSS), Vancouver, BC, July 24–29.

Fensholt, R., K. Rasmussen, T. T. Nielsen, and C. Mbow. 2009. "Evaluation of Earth Observation Based Long Term Vegetation Trends—Intercomparing NDVI Time Series Trend Analysis Consistency of Sahel from AVHRR GIMMS, Terra MODIS and SPOT VGT Data." *Remote Sensing of Environment* 113: 1886–1898. doi:10.1016/j.rse.2009.04.004.

Frey, C. M., C. Kuenzer, and S. Dech. 2012. "Quantitative Comparison of the Operational NOAA-AVHRR LST Product of DLR and the MODIS LST Product V005." *International Journal of Remote Sensing* 33: 7165–7183. doi:10.1080/01431161.2012.699693.

Gutman, G., and J. G. Masek. 2012. "Long-Term Time Series of the Earth's Land-Surface Observations from Space." *International Journal of Remote Sensing* 33: 4700–4719. doi:10.1080/01431161.2011.638341.

Hachem, S., C. R. Duguay, and M. Allard. 2012. "Comparison of MODIS-Derived Land Surface Temperatures with Ground Surface and Air Temperature Measurements in Continuous Permafrost Terrain." *The Cryosphere* 6: 51–69. doi:10.5194/tc-6-51-2012.

He, D., Y. Lu, Z. Li, and S. Li. 2009. "Watercourse Environmental Change in Upper Mekong." In *The Mekong. Biophysical Environment of an International River Basin*, edited by I. C. Campbell, 335–362. New York: Academic Press.

Heidinger, A. K., A. T. Evan, M. J. Foster, and A. Walther. 2012. "A Naive Bayesian Cloud-Detection Scheme Derived from CALIPSO and Applied within PATMOS-x." *Journal of Applied Meteorology and Climatology* 51: 1129–1144. doi:10.1175/JAMC-D-11-02.1.

Holben, B. 1986. "Characteristics of Maximum-Value Composite Images from Temporal AVHRR Data." *International Journal of Remote Sensing* 7 (11): 1417–1434.

Jin, M. 2000. "Interpolation of Surface Radiative Temperature Measured from Polar Orbiting Satellites to a Diurnal Cycle: 2. Cloudy-Pixel Treatment." *Journal of Geophysical Research* 105 (D3): 4061–4076. doi:10.1029/1999JD901088.

Jin, M. L., and R. E. Dickinson. 2010. "Land Surface Skin Temperature Climatology: Benefitting from the Strengths of Satellite Observations." *Environmental Research Letters* 5: 044004. doi:10.1088/1748-9326/5/4/044004.

Kuenzer, C., I. Campbell, M. Roch, P. Leinenkugel, V. Q. Tuan, and S. Dech. 2012. "Understanding the Impact of Hydropower Developments in the Context of Upstream– Downstream Relations in the Mekong River Basin." *Sustainability Science* 11: 1–20.

Kuenzer, C., and S. Dech, eds. 2013. "Theoretical Background of Thermal Infrared Remote Sensing." In *Thermal Infrared Remote Sensing – Sensors, Methods, Applications*. Remote Sensing and Digital Image Processing Series, 1–26. Netherlands: Springer. doi:10.1007/978-94-007-6639-6_1.

Leinenkugel, P., C. Kuenzer, and S. Dech. 2013. "Comparison and Enhancement of MODIS Cloud Mask Products for Southeast Asia." *International Journal of Remote Sensing* 34: 2730–2748. doi:10.1080/01431161.2012.750037.

Leinenkugel, P., C. Kuenzer, N. Oppelt, and S. Dech. 2013. "Characterisation of Land Surface Phenology and Land Cover Based on Moderate Resolution Satellite Data in Cloud Prone Areas —A Novel Product for the Mekong Basin." *Remote Sensing of Environment* 136: 180–198. doi:10.1016/j.rse.2013.05.004.

Moreno-Ruiz, J. A., D. Riano, J. R. Garcia-Lazaro, and S. L. Ustin. 2009. "Intercomparison of AVHRR PAL and LTDR Version 2 Long-Term Datasets for Africa from 1982 to 2000 and its Impact on Mapping Burned Area." *IEEE Geoscience and Remote Sensing Letters* 6: 738–742. doi:10.1109/LGRS.2009.2024436.

Neteler, M. 2010. "Estimating Daily Land Surface Temperatures in Mountainous Environments by Reconstructed MODIS LST Data." *Remote Sensing* 2: 333–351. doi:10.3390/rs1020333.

NOAA. 2013. "Climate Monitoring, Global Surface Temperature Anomalies." Accessed September 25. http://www.ncdc.noaa.gov/cmb-faq/anomalies.php.

Poggio, L., A. Gimona, and I. Brown. 2012. "Spatio-Temporal MODIS EVI Gap Filling Under Cloud Cover: An Example in Scotland." *ISPRS Journal of Photogrammetry and Remote Sensing* 72: 56–72. doi:10.1016/j.isprsjprs.2012.06.003.

Salama, M. S., R. Van der Velde, L. Zhong, Y. Ma, M. Ofwono, and Z. Su. 2012. "Decadal Variations of Land Surface Temperature Anomalies Observed over the Tibetan Plateau by the Special Sensor Microwave Imager (SSM/I) from 1987 to 2008." *Climatic Change* 114 (3–4): 769–781.

Shen, S., and G. G. Leptoukh. 2011. "Estimation of Surface Air Temperature Over Central and Eastern Eurasia from MODIS Land Surface Temperature." *Environmental Research Letters* 6: 045206. doi:10.1088/1748-9326/6/4/045206.

Tian, F., G. Y. Qiu, Y. H. Yang, Y. J. Xiong, and P. Wang. 2012. "Studies on the Relationships between Land Surface Temperature and Environmental Factors in an Inland River Catchment Based on Geographically Weighted Regression and MODIS Data." *IEEE Journal of Selected Topics in Applied Earth Observations and Remote Sensing* 5 (3): 687–698. doi:10.1109/JSTARS.2012.2190978.

Wan, Z., and J. Dozier. 1996. "A Generalized Split-Window Algorithm for Retrieving Land-Surface Temperature from Space." *IEEE Transactions on Geoscience and Remote Sensing* 34: 892–905. doi:10.1109/36.508406.

Westermann, S., M. Langer, and J. Boike. 2011. "Spatial and Temporal Variations of Summer Surface Temperatures of High-Arctic Tundra on Svalbard – Implications for MODIS LST Based Permafrost Monitoring." *Remote Sensing of Environment* 115 (3): 908–922. doi:10.1016/j.rse.2010.11.018.

Xu, Y., Y. Shen, and Z. Wu. 2013. "Spatial and Temporal Variations of Land Surface Temperature over the Tibetan Plateau Based on Harmonic Analysis." *Mountain Research and Development* 33 (1): 85–94. doi:http://dx.doi.org/10.1659/MRD-JOURNAL-D-12-00090.1.

You, Q., S. Kang, N. Pepin, W.-A. Flügel, Y. Yan, H. Behrawan, and J. Huang 2010. "Relationship between Temperature Trend Magnitude, Elevation and Mean Temperature in the Tibetan Plateau from Homogenized Surface Stations and Reanalysis Data." *Global and Planetary Change* 71 (1–2): 124–133.

Sensitivity analysis for predicting continuous fields of tree-cover and fractional land-cover distributions in cloud-prone areas

Patrick Leinenkugel[a], Michel L. Wolters[a], Claudia Kuenzer[a], Natascha Oppelt[b], and Stefan Dech[a]

[a]German Aerospace Center (DLR), German Remote Sensing Data Center (DFD), Oberpfaffenhofen, Germany; [b]Institute for Geography, Christian-Albrechts-Universität zu Kiel, Kiel, Germany

The use of multi-temporal datasets, such as vegetation index time series or phenological metrics, for improved classification and regression performance is well established in the remote-sensing science community. However, the usefulness of such information is less apparent for areas with distinct wet season periods and heavily concentrated cloud cover. In view of this, this study examines the potential of multi-temporal datasets for the estimation of sub-pixel land-cover fractions and percentage tree cover in an area having distinct wet and dry seasons. Prediction is based on a regression tree algorithm in combination with linear least-squares regression planes, which relate multi-spectral and multi-temporal satellite data from the Moderate-Resolution Imaging Spectroradiometer (MODIS) sensor to sub-pixel land-cover proportions and percentage tree cover, derived from high-resolution land-cover maps. Furthermore, several versions of the latter were produced using different classification approaches to evaluate the sensitivity of the response variable on overall prediction accuracy. The results were evaluated according to absolute accuracy levels and according to their long-term inter-annual robustness by applying the regression models to MODIS data over a period of 11 years. The best regression model based on dry season information only estimated continuous fields of percentage tree cover with a prediction error of less than 7% and an inter-annual variability of less than 4% over a time period of 11 years. The inclusion of intra-annual information did not contribute to any improvements in model accuracy compared to information from the dry season alone, and furthermore, deteriorated inter-annual robustness of model predictions. In addition, it has been shown that the quality of the response variable in the training data had significant effects on overall accuracy.

1. Introduction

The distribution of land cover and its dynamics are important variables in global-scale patterns of the climate and biogeochemistry of the Earth system (Matthews 1983; Wilson and Henderson-Sellers 1985; DeFries et al. 1995) as well as for the understanding of human–environmental interactions (Houghton 1994; Lambin et al. 2001). The processes of interest, however, often operate at scales finer than those resolvable with satellites used for regional to global land-cover analyses. Consequently, mixed pixels emerge at the respective scale of interest, i.e. the simultaneous occurrence of multiple land-cover types relative to the size of the sensor's projected instantaneous field of view. It has been shown that by assigning mixed pixels to a specific single type of land cover, inferior mapping performance is generally obtained (Foody et al. 1997; Friedl et al. 2000; Fernandes et al. 2004; Leinenkugel, Kuenzer, Oppelt, et al. 2013) and in particular gives rise to an overall underestimation of non-dominant land-cover types (Nelson and Holben 1986; Braswell et al. 2003). In the past decades, various techniques have been developed that exploit more effectively the information inherent in the spectral signature of each pixel, thereby aiming at the quantification of land-cover characteristics or biophysical properties at sub-pixel level. The derived variables are of continuous nature and thus are generally better equipped for the characterization of land-cover gradients and mosaics in the landscape (Adams et al. 1995; DeFries, Hansen, and Townshend 2000). Such methods include linear mixture models (Adams et al. 1995; DeFries, Hansen, and Townshend 2000; Scanlon et al. 2002; Lu, Moran, and Batistella 2003; Kuenzer et al. 2008; Kumar et al. 2008), fuzzy membership functions (Foody and Cox 1994), neural networks (Foody et al. 1997; Carpenter et al. 1999; Braswell et al. 2003; Liu et al. 2004; Liu and Wu 2005), support vector machines (Esch et al. 2009; Leinenkugel, Esch, and Kuenzer 2011), and regression trees (DeFries et al. 1997; DeFries, Townshend, and Hansen 1999; Hansen et al. 2002; Gessner et al. 2013). A number of studies focused on the analytical comparison of various unmixing and regression techniques as performed by Fernandes et al. (2004), who compared the result from a conventional discrete classifier with an artificial neural network, a clustering/lookup-table approach, multivariate regression, and linear least-squares inversion for mapping sub-pixel land-cover fractions and continuous fields of vegetation characteristics. Liu and Wu (2005) compared four non-linear regression models, i.e. an adaptive resonance theory map (ARTMAP), an adaptive resonance theory mixture map (ART-MMAP), a regression tree algorithm, and a multilayer perceptron for sub-pixel land-cover classification. Schwarz and Zimmermann (2005) derived continuous fields of percentage tree cover with a generalized linear model and compared the results with those obtained from a regression tree algorithm. However, only a few studies have extended the analyses beyond comparing different unmixing or regression algorithms to include other parameters that influence the accuracy of sub-pixel estimates. Braswell et al. (2003), for instance, evaluated the effect of atmospheric correction and view angle effects on model accuracy for a Bayesian-regularized artificial neural network. Townshend et al. (2000) carried out a simulation experiment to analyse the effects of the modulation transfer function on estimating sub-pixel land-cover proportions by linear mixture modelling. Shao and Lunetta (2011) examined sub-pixel land-cover classification performance for an area-wide approach against an ecoregion-based approach and also incorporated the effects of seasonality in their experiments. Most studies based on moderate-resolution satellite data include temporal reflectance composites (Braswell et al. 2003; Hansen et al. 2005; Guerschman et al. 2009), vegetation index time series (Tottrup et al. 2007), or phenological metrics (DeFries, Townshend, and Hansen 1999; Schwarz and Zimmermann 2005) in the explanatory variables for model construction. Although for many regions the influence of cloud cover may not be a limiting factor, certain regions are, however, characterized by distinct wet seasons that hamper the use of optical image data throughout the year. Although methods exist that reduce the variability related to clouds, haze, and atmospheric contamination, the processing may have significant effects on the spectral and spatial fidelity of the input data (Hansen et al. 2005).

Although for global approaches the incorporation of intra-annual datasets has been seen to improve accuracy in classification or regression exercises (DeFries, Hansen, and Townshend 1995), this improvement, however, is less apparent for areas with periods of heavily concentrated cloud cover. In this context, this study examines the potential of multi-temporal Moderate-Resolution Imaging Spectroradiometer (MODIS) data for the estimation of sub-pixel land-cover fractions and percentage tree cover for an area with a distinct wet season period, through the inclusion of different composite types, vegetation index time series, and phenological metrics in the explanatory dataset. Furthermore, in addition to conventional accuracy metrics, inter-annual robustness of the output predictions is analysed by applying the regression models to satellite data over a period of 11 years. In this way, the potential of the different explanatory variables for estimating land-cover change over long-term data records can be assessed. Finally, no study to date has investigated the effects of the integrity of the response variable on the accuracy of the final prediction outputs. For this purpose, different classification algorithms were applied to different input feature datasets in the process of training data generation to evaluate the sensitivity of these variables on model prediction.

2. Study area and data

This study focuses on the lower Mekong Basin covering large parts of Laos, Thailand, Cambodia, and some smaller areas of Myanmar and Vietnam (Figure 1). The region, spanning a total area of approximately 620,000 km^2, is characterized by a moderate terrain and includes various land-cover types ranging from semi-evergreen and dry-deciduous broadleaved forests and woodlands to intensive cultivated planes of cropland with up to three harvest cycles per year. The summer southwest and winter northeast monsoons are responsible for a distinct bi-seasonal pattern of wet and dry periods. This is manifest in a period of 2–5 cool, dry months between November and April with precipitation rates below 60 mm per month, followed by a hotter and wetter period lasting from May to October, with precipitation rates up to 500 mm per month.

Within the lower Mekong Basin, three Landsat tiles, together covering a region of approximately 185 km × 500 km, were selected as calibration and validation areas. For each of the three tiles, Landsat ETM+ data for the year 2000 and Landsat TM data for the year 2010 were chosen from the United States Geological Survey (USGS) archive, in such a way that each scene was characterized by relatively low cloud cover and that the scenes for the respective years were acquired within the shortest possible time interval (see Table 1).

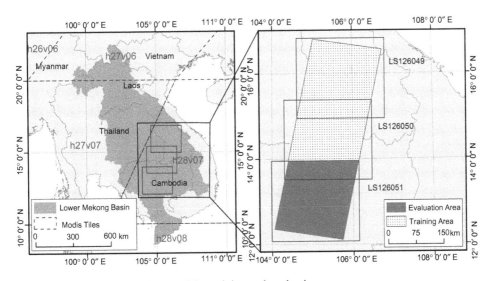

Figure 1.   Study area and data used for training and evaluation.

Table 1. Landsat data used in this study.

Scene	Path/row	Acquisition date	Cloud cover (%)
LE71260492000334EDC00	126/49	29 November 2000	5
LE71260502000366SGS01	126/50	31 December 2000	0
LE71260502000366SGS01	126/51	31 December 2000	0
LT51260492010033BKT00	126/49	2 February 2010	1
LT51260502010033BKT00	126/50	2 February 2010	2
LT51260502010033BKT00	126/51	2 February 2010	2

Furthermore, satellite data from the MODIS instrument aboard the Terra and Aqua satellites were used. The standard MOD09A1 eight day composite products were selected as base data providing surface spectral reflectance in the MODIS bands 1–7 (620–670 nm, 841–876 nm, 459–479 nm, 545–565 nm, 1230–1250 nm, 1628–1652 nm, and 2105–2155 nm) at 500 m spatial resolution, these being corrected for the effects of atmospheric gases and aerosols (Vermote, El Saleous, and Justice 2002). For the time period 2001–2011, the respective composites from both platforms, Aqua and Terra, were acquired for the MODIS tiles h27v06, h27v07, h28v07, and h28v08, from the Land Processing Distributed Active Archive Center (LP DAAC), which together cover the lower part of the Mekong Basin area.

3. Methods

3.1. Generating training data

3.1.1. Generation of response variables

The Landsat scenes, processed to Standard Terrain Correction (L1T), were used to generate the response variables, i.e. fractional land-cover and percentage tree-cover estimates, for the years 2001–2011 (Figure 2). To account for radiometric distortions related to the atmosphere, each Landsat scene was atmospherically corrected using ATCOR-3 (Richter 1997), which also incorporates the correction of topography-related biases through the inclusion of information on elevation and slope from the Shuttle Radar Topography Mission (SRTM). Since the respective scenes showed a very accurate spatial coincidence with x- and y-ground control point (GCP) residuals below 9 m (obtained from the respective GCP Residual Report metafile), no additional geometric correction was required.

Areas related to land-cover changes between 2001 and 2010 were identified by relative change detection utilizing a spectral angle mapping (SAM) approach (Kruse et al. 1993). Thereby, the spectral angle between corresponding pixel spectra in the images is computed, indicating the spectral similarity between two pixels within the feature space. Subsequently, a threshold value was defined empirically that separates spectral differences related to noise and insignificant changes on the land surface (e.g. illumination) from those related to significant land-cover and land-use changes. Clouds that have not already been identified by the change detection method were excluded by applying the Automatic Cloud Cover Assessment (ACCA) procedure (Irish et al. 2006). After all clouds and areas of land-cover change were identified, the three Landsat TM scenes from 2010 were classified into five land-cover classes – evergreen broadleaved forest, broadleaved woodland, cropland, herbaceous vegetation, and water. To test the effect of different response variables on the accuracy of model predictions, six different

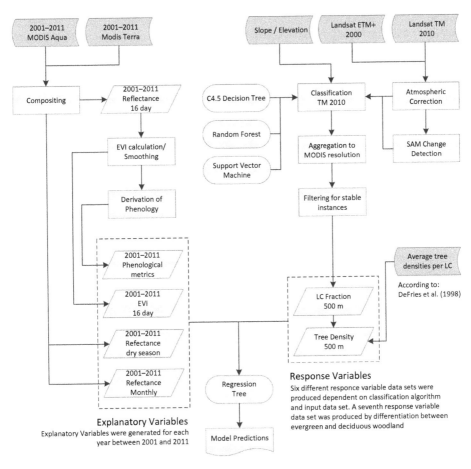

Figure 2.   Overview of the input feature processing and classification/regression procedure.

land-cover maps were generated by applying three different classification algorithms – Random Forests (Breiman 2001), C4.5 (Quinlan 1993), and Support Vector Machines (Cortes and Vapnik 1995) – on two different attribute sets. The latter constituted the seven spectral Landsat bands and the spectral bands in combination with information on elevation and slope from STRM data. The inclusion of slope and elevation aimed to examine whether mapping accuracy can be increased due to the strong relation between land cover and topography in the study area, as experienced in previous studies (Lu and Weng 2007). Furthermore, a sixth land-cover map was generated based on Support Vector Machine classification that additionally separated evergreen and dry-deciduous woodland, as the satellite images were acquired in the dry-season period. One concern in any comparative assessment is the possibility that the classification algorithms were unequally tuned to the data to be classified. In this respect, human interaction was minimized by setting the parameters of the respective classifiers by maximizing the cross-validation accuracy of the training data. Thereby, all relevant parameters for a certain algorithm were tested systematically within a predefined parameter range and by a predefined step-size.

Finally, the parameter combination with the minimum training error was selected. Training areas for classification were defined manually on the basis of Google Maps imagery and field photographs of the Global Geo-Referenced Field Photo Library at the University of Oklahoma (Xiao et al. 2011), which were compared with the actual Landsat scene. Furthermore, each Landsat scene was initially clustered into 30 spectral classes to facilitate the visual differentiation of the land-cover types on the ground and to ensure that spectrally homogeneous training samples were defined. A total of 939 training polygons comprising 100,594 training pixels were collected distributed over all three Landsat scenes. The number of training instances reached from 11,702 pixels (234 samples) for herbaceous vegetation to 41,014 pixels (410 samples) for the woodland class. For valida-tion purposes an independent test set was defined by randomly distributing a total of 971 sample points stratified according to the respective land-cover proportions. Subsequently, for each point the respective land-cover type on the ground was evaluated based on very-high-resolution imagery from Google Earth.

The thematic land-cover maps were re-projected from UTM map projection to geo-graphical coordinates using nearest neighbour re-sampling. Finally, each land-cover class was labelled with a mean tree-cover estimate (Table 2) according to DeFries et al. (1998), and was aggregated to derive fractional land-cover and continuous fields of percentage tree cover at a spatial resolution of 500 m. To ensure that areas of land-cover change are excluded from further analyses, aggregated pixels were filtered and only instances con-taining less than 3% of detected change were retained (Figure 3). The resulting fractions and tree-cover estimates are unevenly distributed, indicating the degree of spatial homo-geneity of the respective land-cover classes (Figure 4). Heterogenic land-cover types such as herbaceous vegetation follow a distinct Poisson-like distribution with very high numbers of pixels with low fractions and rapidly decreasing numbers towards higher fractions.

3.1.2. Generation of MODIS explanatory variables

To reduce biases related to residual cloud cover and haze, cloud shadows, sensor defects, and surface anisotropy, the composite length of eight days for the standard MODIS composites was extended, thereby combining suitable observations from the platforms Aqua and Terra. First, 16 day composites were generated based on the Constrained-View angle – Maximum NDVI Value Composite (CV-MNVC) algorithm (Huete et al. 2002) and were subsequently used to calculate a vegetation index time series. Furthermore,

Table 2. Definitions of land-cover types and mean canopy cover estimates according to DeFries et al. (1998), defined as the percentage of ground surface area covered by a vertical projection of tree crowns (White, Shaw, and Ramsey 2005).

	Tree cover (%)	Tree cover evergreen (%)	Tree cover deciduous (%)
Croplands	0	00.0	00.0
Herbaceous vegetation	25	12.5	12.5
Deciduous broadleaf woodland	50	10.0	40.0
Evergreen broadleaf woodland	50	40.0	10.0
Evergreen broadleaf forests	80	80.0	00.0
Water	0	00.0	00.0

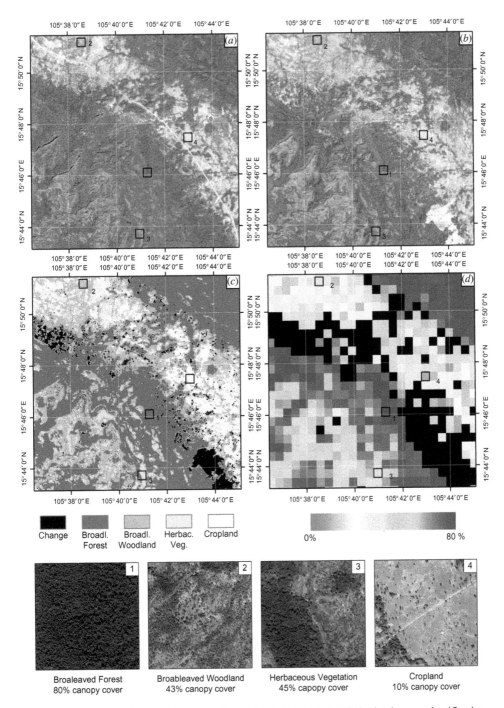

Figure 3. Subset for the Landsat scenes from (*a*) 2000, (*b*) 2010, (*c*) the land-cover classification as derived from the 2010 scene, and (*d*) derived tree-cover estimates. In addition, high-resolution Google Earth imagery for four different MODIS pixels is shown with the extent indicated in the subsets above. Each plot is primarily characterized by one specific land-cover type as derived from the Landsat scenes with the resulting canopy cover estimates stated below.

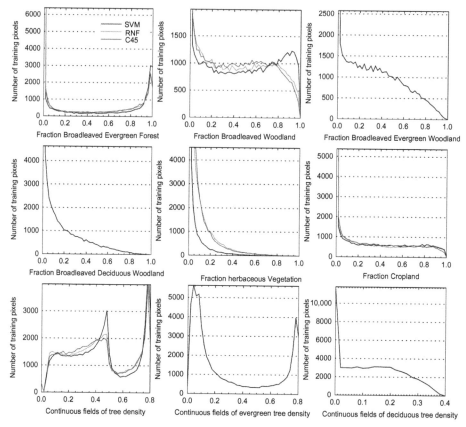

Figure 4. Distribution of fractions for the respective land-cover types for classification results derived with a Support Vector Machine (SVM), Random Forests (RF), and C4.5. Differentiation according to evergreen and deciduous woodland was only applied in the case of SVM classification. The x-axis for a respective biome is defined by its maximum valid range and the y-axis is limited to a fifth of the global maximum occurrence.

monthly and dry-season composites, including all observations between November and February, were generated for all spectral bands excluding band 5 (1230–1250 nm), based on the Mean Value Composite (MVC) method (Vancutsem et al. 2007). Band 5 was excluded from further processing, as significant sensor defects were evident in the composites for some of the years. For the CV-MNVC and MVC methods, all observations labelled as cloudy, adjacent to cloud, or showing view angles above 60° were removed. Furthermore, an additional empirically defined threshold in the blue band of 0.09 was introduced to ensure that residual clouds and observations with high aerosol concentrations not correctly identified by the cloud masks were excluded. Finally, the mean value was computed for each pixel based on the remaining observations for the MVC products whereas for the CV-MNVC composites observation with the lowest view angle and highest normalized difference vegetation index (NDVI) was computed.

Based on the 16 day composites, the enhanced vegetation index (EVI) was calculated from the blue, red, and NIR bands according to Huete et al. (2002). Data gaps and high-frequent noise in the EVI time series were effectively reduced by applying an adaptive Savitzky–Golay filter. Since the cropland area in the study area is characterized by up to three harvest cycles per year, different filter window sizes were necessary to preserve the phenological characteristics of the land surface (Leinenkugel, Kuenzer, Oppelt, et al. 2013). Hence, different parameterizations were applied depending on the number of harvest cycles for each pixel, which were defined by an initial harmonic analysis, based on an EVI time series of five years. The detailed methodology was previously described by Leinenkugel, Kuenzer, Oppelt, et al. (2013) and therefore will not be repeated in this study. Phenological metrics (minimum, maximum, mean, and range) were derived from the smoothed EVI time series according to Hansen et al. (2002). Thereby metrics are computed for the composites with the 23, 16, 12, 8, and 4 highest EVI values, respectively, resulting in 20 metrics per pixel that represent the phenological characteristics of the land surface. It has been shown that by selecting only the months with the highest EVI, the effect of varying seasonality across the study area and the influence of atmospheric contamination on the time series can be minimized (Hansen et al. 2002; Schwarz and Zimmermann 2005).

3.2. Numeric prediction using a regression tree

In this study the model tree algorithm M5 (Quinlan 1992), as implemented in the Waikato Environment for Knowledge Analysis (WEKA) (Hall et al. 2009), was used to generate the annual fractional land-cover and percentage tree-cover estimates for the years 2001–2011. Model trees belong to the family of regression trees (Breiman et al. 1984) that have been frequently used in remote sensing for numeric prediction and fractional land-cover estimation (Defries et al. et al. 1997; DeFries, Townshend, and Hansen 1999; Hansen et al. 2002, 2005; Tottrup et al. 2007; Gessner et al. 2013). A regression tree is constructed by binary recursive partitioning whereby a dataset is successively split into progressively homogeneous subsets (Schwarz and Zimmermann 2005). For each split or node, a splitting criterion is used to determine which explanatory variable is most suited to split the training data. In this implementation, the explanatory attribute is chosen that minimizes the intra-subset standard deviation at a node (Witten, Eibe, and Hall 2011). The splitting process terminates when the standard deviation at a node is only less than 5% of the standard deviation of the original instance set or when less than 10 instances remain at a node. Once the tree has been constructed, it is pruned back from the leaves, in order to counter the effect of over-fitting. This is carried out as long as the expected error at a node remains lower than the expected error from the sub-tree below. In contrast to ordinary regression trees, a linear regression model is constructed at each terminal node or leaf that predicts the value of the instances that reach the leaf (Wang and Witten 1996). Thereby, an additional smoothing process is implemented to reduce sharp discontinuities between adjacent linear models at the leaves of the pruned tree Additional theoretical background on the model tree implementation within WEKA can be found in Wang and Witten (1996) or Witten, Eibe, and Hall (2011).

3.3. Experimental design

The accuracy of predicting land-cover fractions and continuous fields of tree cover was tested under various combinations of training datasets relative to a base case (see Table 3). The experimental design was developed to provide a better understanding of the effects of different spectral and temporal resolutions on model accuracy for areas characterized by a distinct wet and dry season. Thereby different combinations of the monthly and dry-season spectral composites (M-SB123467 and DS-SB123467), 16 day EVI composites

Table 3. Tested scenarios for fractions of forest, woodland, herbaceous vegetation, and cropland, as well as for continuous fields of tree density.

Analysis	Explanatory variable	Response variable	
		Classifier	Data set
Influence of explanatory attribute test sets on prediction accuracy	DS-SB123467* DS-SB123467 + Pheno M-SB123467 all year M-SB123467 dry season 16-EVI all year Pheno	SVM*	TM-SB*
Influence of seasonality on prediction accuracy	M-SB123467 single composite 16-EVI single composite	SVM	TM-SB
Influence of response attribute test sets on prediction accuracy	DS-SB123467	SVM RF C4.5	TM-SB + ES
		SVM RF C4.5	TM-SB + ES

Note: *Base case.

(16-EVI), and the phenological metrics (Pheno) were tested for different periods throughout the year. In this context, an auxiliary dataset on daily cloud-cover proportions for the study area was computed from the daily MODIS MOD09GA surface reflectance product at 1 km resolution for the year 2010. Furthermore, different versions of the responsive variable were tested, derived from classifications by a Support Vector Machine (SVM), Random Forests (RF), and C4.5. The classifiers were applied on two datasets – the Landsat Thematic Mapper spectral bands (TM-SB) and the Landsat Thematic Mapper spectral bands including elevation and slope (TM-SB + ES).

Since model performance generally deteriorates for spatially distant instances compared to those close to the training samples, the model results were evaluated on a distant and a near test set in relation to the training data. Therefore, every third training pixel in the scenes LS126049 and LS126050 was excluded from training and formed the proximate test set, whereas the distant test set comprised all pixels from the scene LS126051 (Figure 1). The predicted tree-cover estimates against the test data were evaluated based on three different statistical measures – the mean error (ME), the root-mean-square error (RMSE), and the coefficient of determination (R^2). Furthermore, the inter-annual robustness of the results was evaluated by the per-pixel standard deviation of 11 years of model predictions. Since the evaluation of the standard deviation does not require any test data, the validation area was extended to the area of the lower Mekong Basin.

4. Results

4.1. Prediction accuracy for the base case

Deviations between predicted land-cover fractions and tree-cover estimates compared to the test data varied to a great extent over the valid range, lying between 0–100% and 0–80%, respectively. Figure 5 illustrates for each model the RMSE and ME statistics for the base case as a function of land-cover fraction and tree-cover estimate, as derived from the near test set for 2010. Furthermore, the global average RMSE and ME are indicated in the

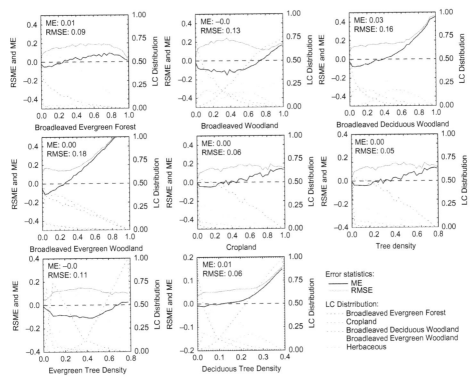

Figure 5. RMSE and ME for predicted land-cover proportions and continuous fields of percentage tree cover (solid lines) for 2010 in 3% intervals. For each interval, average land-cover composition for the test data is indicated by the dotted lines. The *y*-axis is limited to half of the valid range.

top-left corner of each plot. When not differentiating between evergreen and deciduous woodland, the RMSE never exceeds a level of 0.25. The global RMSE and ME for each model are comparatively low, lying between 0.05 and 0.13 owing to the high numbers of pixels with low fractions, which generally show very high levels of accuracy. All models overestimate low and intermediate fractions and underestimate pixels with high land-cover proportions, as indicated by the ME. This pattern is particularly evident for the woodland class, whereas for evergreen broadleaved forests the ME balances out towards pure pixels. Accuracy becomes significantly less when differentiating between evergreen and deciduous woodland. Both show significant increases in RMSE towards the high fractions, which can be almost exclusively attributed to underestimations. The same pattern could also be observed for the RMSE and ME of predicted continuous fields of deciduous tree cover with a maximum underestimation of 0.15. Generally, however, predicted tree-cover estimates showed significantly higher levels of accuracy compared to those for land-cover fractions. Evergreen tree cover is generally overestimated with a maximum RMSE of 0.17 in the intermediate range where the proportion of evergreen woodland is the highest. When not differentiating tree cover according to seasonality, deviations generally remain low with an RMSE not exceeding 0.1. Herbaceous land cover in the study area is distributed in a very heterogenic pattern of small patches of grass and shrubland. Consequently, insufficient pixels with high fractions of herbaceous vegetation were available in the test and training data and, thus, results for this class are not shown separately here.

4.2. The effect of explanatory attribute test sets on prediction accuracy

Figure 6 illustrates model performance for predicting continuous fields of tree cover in relation to cloud-cover conditions in the study area throughout the year 2010. The figure shows the RMSE and R^2 values for the distant test set derived from individual models with single-date composites as input features. During the months of the dry season, between November and March, daily cloud-cover proportions of about 40% and RMSE values of 0.06–0.10 for the respective models are the lowest. Cloud cover and RMSE values, though, rapidly increase in the wet season between May and October. During this time daily cloud cover increases up to 95% in the study area while the monthly RMSE rises up to 0.25 for models based on the reflectance composites and up to 0.17 for those based on the 16 day EVI composites. The seasonal effect on model performance is even more pronounced when evaluating R^2, showing an annual range of 0.52 for the monthly spectral bands and 0.43 for the EVI composites. In the dry season, the monthly reflectance composites resulted in a significantly better model performance compared to the individual 16 day EVI composites. In the wet season, however, the EVI time series show a temporally more coherent model performance and leads to a performance improvement relative to the monthly reflectance composites. Furthermore, for the EVI composites, a small shift between the trajectory of the error statistics and the cloud-cover distribution through the year can be observed.

Although accuracy levels for the wet season are significantly lower than for the dry season, the application of the entire annual time series of 16 day EVI composites or the

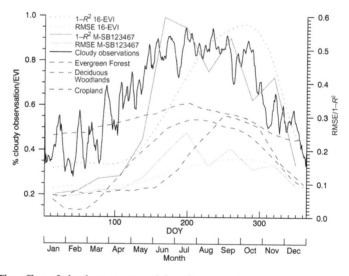

Figure 6. The effect of cloud cover on model performance for predicting percentage tree cover. Clear-sky proportions are obtained from daily MOD09 MODIS data covering the lower Mekong Basin.

monthly spectral composites did not deteriorate the overall accuracy as indicated by the 11 year RMSE and R^2 (Table 4). On the other hand, the phenological information inherent in the time series also did not contribute towards an improved differentiation of the dense evergreen forests and the less-dense deciduous woodlands. Furthermore, the inclusion of phenological metrics generally had negative effects on inter-annual stability, as indicated by the 11 year standard deviation of predicted tree-cover estimates (see Figure 7). The differentiation according to various land-cover types based on the Mekong LC2010 product (Leinenkugel, Kuenzer, Oppelt, et al. 2013) reveals that the effects of different explanatory datasets on the intra-annual variability is particularly distinctive for the forest and the woodland classes. It can be seen that the exclusive use of the dry-season

Table 4. 11 year average of root mean square error (RMSE) and coefficient of determination (R^2) for different test cases as derived from the near and distant test data.

	11 year average	RMSE near	RMSE distant	R^2 near	R^2 distant
Explanatory datasets	DS-B123467	0.05	0.07	0.94	0.92
	DS-B123467 + Pheno	0.05	0.07	0.94	0.90
	M-B123467 all year	0.05	0.07	0.96	0.92
	M-B123467 dry season	0.05	0.07	0.96	0.92
	16-EVI all year	0.07	0.09	0.90	0.88
	Pheno	0.07	0.08	0.92	0.88
Response datasets	SVM on TM-SB	0.05	0.07	0.94	0.92
	RF on TM-SB	0.07	0.08	0.94	0.88
	C4.5 on TM-SB	0.06	0.07	0.94	0.90
	SVM on TM-SB + ES	0.06	0.07	0.94	0.92
	RF on TM-SB + ES	0.09	0.15	0.90	0.67
	C4.5 on TM-SB + ES	0.08	0.14	0.90	0.74

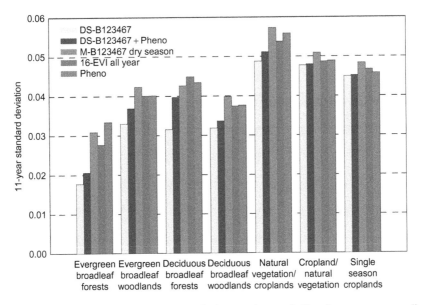

Figure 7. 11 year standard deviation for predicting continuous fields of tree cover according to land-cover types derived from the Mekong LC2010 product (Leinenkugel, Kuenzer, Oppelt, et al. 2013).

composites showed the lowest inter-annual variation of all test sets and increased up to ~ 1.5% when including additional temporal information from the wet season. The reduction of the composite length from dry-season composites to monthly composites within the dry season also showed no significant effect on the accuracy levels but remarkable effects on inter-annual variability. The exclusive use of the EVI time series or the phenological metrics achieved satisfactory results in absolute model performance, with only 0.02 lower RMSE and R^2 relative to the base case, however, resulted among the highest inter-annual variability, particularly distinctive for the forest and woodland classes.

4.3. The effect of response attribute test sets on prediction accuracy

Furthermore, the relevance of quality differences in the response variable was tested by producing six different high-resolution land-cover maps that were used for creating training fractions. Table 5 shows the validation results for the respective land-cover maps. Overall accuracy ranged from 82% for the SVM classification on the spectral bands to 73% for the random forest and C4.5 classification based on the spectral bands in combination with slope and elevation. The inclusion of slope and elevation deteriorated classification accuracy for all classifiers. This was particularly evident for the decision-tree results, which showed frequent confusions of woodlands and forests within the LS126051 scene. The quality differences between the land-cover maps are also reflected in the accuracy levels of the predicted tree-cover estimates. The lowest RMSE of 0.05 and the highest R^2 of 0.94 for the near test set were achieved by the SVM on the spectral bands only, with slightly lower accuracy levels for the distant test set. Corresponding to the Landsat classification results, accuracy levels for the predicted tree-cover estimates decreased when the random forest and C4.5 training datasets were used for model calibration. This was particularly evident for the distant test and when information on slope and elevation was included in the classification process. Furthermore, it is evident from Figure 8 that particularly large differences in prediction accuracy could be observed for the lower fraction ranges of broadleaved evergreen forests and in the high ranges of broadleaved woodland and cropland.

Table 5. Producer's accuracy (PA), user's accuracy (UA), overall accuracy (OA), and kappa coefficient (κ) for the different high-resolution land-cover maps used for creating training fractions.

			Water	Forest	Cropland	Woodland	Herbaceous		
SVM on TM-SB	PA		1	0.86	0.79	0.89	0.52	OA	0.82
	UA		1	0.97	0.84	0.69	0.79	κ	0.75
RF on TM-SB	PA		1	0.82	0.84	0.79	0.52	OA	0.79
	UA		1	0.94	0.81	0.68	0.60	κ	0.71
C4.5 on TM-SB	PA		1	0.82	0.80	0.79	0.47	OA	0.77
	UA		0.95	0.94	0.81	0.67	0.54	κ	0.69
SVM on TM-SB + ES	PA		1	0.74	0.79	0.87	0.51	OA	0.77
	UA		1	0.93	0.86	0.62	0.75	κ	0.69
RF on TM-SB + ES	PA		1	0.72	0.77	0.77	0.53	OA	0.73
	UA		1	0.92	0.86	0.60	0.48	κ	0.63
C4.5 on TM-SB + ES	PA		1	0.71	0.76	0.78	0.50	OA	0.73
	UA		0.95	0.92	0.85	0.59	0.50	κ	0.63

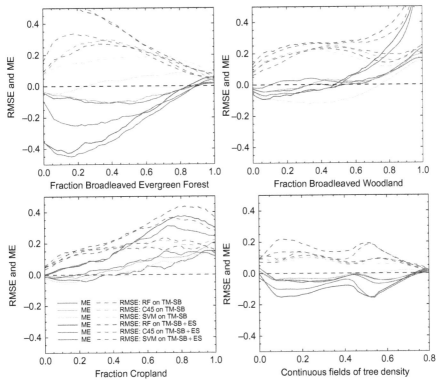

Figure 8. RMSE and ME statistics for continuous fields of tree cover of 2010 as derived for the base case and the distant test set. The different error statistics result from different response datasets used in the process of model calibration.

5. Discussion of the results

5.1. On the prediction of fractional land-cover and percentage tree cover

The accuracy for the different land-cover types varies significantly along the entire continuum, and is closely linked to the respective land-cover composition within a pixel and the distribution of training pixels across the valid range. Unmixing techniques generally assume linear relationships between the sub-pixel land-cover proportions under the prerequisite that the respective pure land-cover types are spectrally independent. This assumption is, however, to some degree violated for classes with spectrally similar characteristics. The regression tree algorithm is an advancement in this respect, since it can manage non-linear relationships without making assumptions about the physics of sub-pixel mixing (Braswell et al. 2003; Hansen et al. 2002). Furthermore, heavily skewed, unimodal distributions of training instances may cause serious over- or underestimations in the output predictions. The high fractions of evergreen woodland, for instance, are heavily underestimated as a result of their spectral proximity to evergreen broadleaved forest and deciduous woodland that together constitute the majority of training pixels. It was seen that spectral overlap in the feature space was reduced and the proportion of training pixels particularly for the high fractions was increased, when not differentiating between evergreen and deciduous woodland (Figure 3). As a result, underestimation of pure woodland pixels turned out to be far lower when not differentiating according to seasonality. Nevertheless, the estimation of land-cover compositions for biomes with similar spectral characteristics generally results in lower mapping accuracy and therefore such estimates should only be considered together with information about their uncertainties. The higher accuracy levels for the tree-cover estimates were expected, given the fact that the biophysical and spectral similarity between certain land-cover types such as between the forest and the woodland classes was addressed by weighting each land-cover type with quasi-continuous biophysical characteristics instead of using Boolean class values. In this way, after aggregation to MODIS resolution, correlation between explanatory and response training data becomes higher, resulting in more accurate regression models.

5.2. On the effect of phenology and cloud cover on prediction accuracy

Several tests were performed to evaluate the contribution of multi-temporal information on the accuracy of model predictions. It has been shown that model performance of the single-date models significantly decreased towards the wet season. This decrease can be largely attributed to the significant quality deterioration of the monthly composites from residual clouds, cloud shadows, and haze, as shown in Figure 9. Since the 16 day EVI composites were additionally smoothed to reduce the effects of residual clouds, cloud shadows, and haze, less noise and fewer artefacts were evident in the input data. This resulted in a less-severe decline and higher temporal coherence in model performance during the wet season compared with the monthly composites. In addition, differing

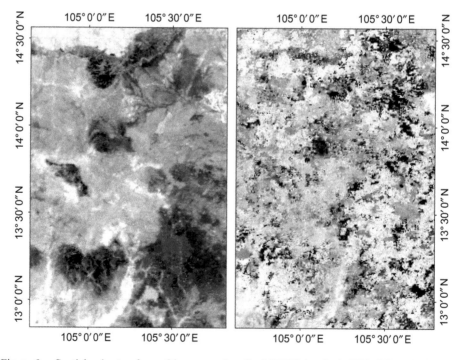

Figure 9. Spatial subsets of monthly composites for MODIS bands 1 (620–670 nm), 3 (459–479 nm), and 4 (545–565 nm) for January (*a*) and August (*b*) representing the dry season and rainy season, respectively.

phenological states in the vegetation cover affect the potential of differentiating between the various land-cover types. Particularly between September and October, the differences in EVI levels amongst single-season cropland, woodland, and evergreen broadleaved forest decrease significantly, as can be observed from the example EVI time series provided in Figure 6. For the monthly spectral composites, the effect of cloud cover on model accuracy can be expected to be more severe than that of a reduced spectral differentiation in the late wet season, whereas for the smoothed EVI composites the conditions are reversed. Consequently, for the monthly spectral composites the lowest accuracy levels can be observed during the core wet season between June and September and for the EVI composites the time point of maximum RMSE and R^2 is shifted towards the end of the wet season between September and October. It has to be noted, however, that next to the above-discussed effects, a certain amount of deterioration in model performance for the wet season composites may be attributed to the fact that the responsive variables were exclusively derived from the dry-season imagery. It can be assumed, however, that classification results may differ depending on the acquisition period of the imagery used for classification. In this respect the dry-season land-cover map can be expected to show a naturally higher correspondence to the dry-season composites than to the monthly composites from the growing season.

Generally, however, the influence of the different explanatory test sets on overall model accuracy has been shown to be less distinctive. It seems that the disruptive external effects of clouds, haze, and cloud shadows, as seen in the wet season, may counteract any improvements from multi-temporal datasets throughout the year. Another further consideration is that prediction results are very high with an RMSE of around 7% and R^2 of 0.93, leaving little potential for improvement. Schwarz and Zimmermann (2005), for instance, also achieved very high accuracies with a mean absolute error of 9% for predicted tree cover in the European Alps. Furthermore, their results deteriorated slightly when all monthly composites within a year were included compared to using only those for the best 4–6 months. Shao and Lunetta (2011), however, experienced that the use of the entire time series of daily MODIS images significantly increased the accuracy of predicted tree canopy up to 10% for a study area in Canada when compared with the results of the best single-date images. Their best results, however, only reached an R^2 of 0.57.

Furthermore, it has been shown that increasing temporal resolution or including phenological information from the wet season has negative effects on the inter-annual robustness of model predictions. Particularly for the western and northern mountain areas in the Basin predominately covered by evergreen forests and woodland, very long periods of consecutive cloud cover occur, which last too long for monthly composite lengths. As a result, monthly composites as well as the EVI time series and the derived phenological parameters showed, despite the laborious processing, rather high variances at pixel level in these regions. As can be observed from Figure 10, the highest variances in the EVI time series, as expressed by the absolute deviation from the 11 year mean, can be observed in the corresponding forest and woodland regions mentioned above (when ignoring the seasonal inundated regions in the southern part of the Basin). The less-cloudy and rather plain cropland areas in contrast show relatively low variances. This pattern corresponds well with the fact that for the forest and the woodland classes, the application of the EVI time series and phenological information in the model resulted in significantly higher inter-annual variability in the output predictions compared to the use of the dry-season information, whereas for croplands, inter-annual variability was almost at the same level for all explanatory datasets (Figure 7).

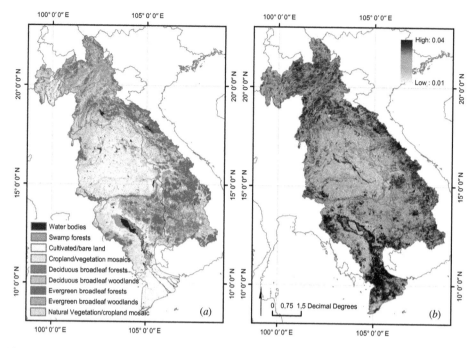

Figure 10. (*a*) Land-cover distribution and variability of the EVI time series, (*b*) expressed as absolute deviation from the 11 year mean for the lower Mekong Basin.

In addition, the effects of surface and atmospheric reflectance anisotropy, known as bi-directional reflection distribution function (BRDF) effects, can be expected to be far lower when averaging the observations over longer time periods. Although a respective MODIS product has been developed to correct for BRDF effects (Schaaf et al. 2002), it has not been incorporated in this study as it is generally unsuitable for this geographical region (Leinenkugel, Kuenzer, and Dech 2013). According to these findings, the sole use of the dry-season composites incorporating four times more observations than the monthly observations seems to offer the best choice for multi-temporal analyses over long-time data records.

5.3. On the quality of the response variable on prediction accuracy

Nearly all studies for generating fractional land cover or continuous fields of biophysical variables are based on training data, which was in turn based on classified high-resolution satellite data (Hansen et al. 2002; Tottrup et al. 2007; Leinenkugel, Esch, and Kuenzer 2011; Gessner et al. 2013). Even at high spatial resolution the differentiation of various vegetation biomes is not straightforward and becomes increasingly difficult when classi-fying multiple scenes from different acquisition dates. Although an overall classification accuracy of 81% for the SVM classification can be regarded as satisfactory, it could be shown that the use of different classification algorithms and the inclusion of slope and elevation resulted in accuracy decreases of up to 10%. The inferior classification perfor-mance when including slope and elevation may be explained by the rather moderate terrain for which the relation between land-cover distribution and topography is less pronounced or often spurious. Furthermore, particularly, decision-tree classifiers are very sensitive to non-representative training data, which could have been the case with respect to slope and elevation within the LS126051 scene. The latter was the basis for the distant test set and showed the largest differences in classification performance. Here, confusions of forests and woodlands were evident for the random forest and the C4.5 classifiers, which, however, were correctly mapped by the SVM algorithm. This high-lights the fact that for optimal classification results, training data collection ideally should be adjusted to the classification algorithm. Since the study only used one training dataset for all classifiers, a comparative conclusion on the general classification capacity of each algorithm should therefore not be drawn from these results.

The quality of the different land-cover maps further propagated into the model-building process as demonstrated by the general highest prediction accuracy for the SVM map with decreasing accuracies for the remaining algorithms. A maximum classi-fication accuracy difference of 9% resulted in a maximum RMSE difference of 4% for the predicted tree-cover estimates, as derived for the near test set. This is twice as high as the maximum RMSE difference related to the various explanatory datasets tested. The larger RMSE difference in the distant test set, however, was not a consequence of low-quality calibration data, but was mainly a result of classification errors in the LS126051 scene. The distant test set in this study should therefore be considered with caution for evaluating the quality of the response variable on regression performance.

This study follows the methodology of Hansen et al. (2002) by labelling each high-resolution land-cover type with an average tree-cover estimate, before aggregating the training data to MODIS resolution. This implies, however, that at high resolution, the woodland class is uniformly assigned a mean canopy cover of 50%, although canopy cover of woodlands in reality may vary locally between 40 and 60%. Consequently, the discretization of a continuous variable into four hard strata inevitably obscures hetero-geneity in vegetation characteristics, which may introduce some additional uncertainties to the training and validation data. However, the derivation of more accurate and spatially explicit tree-cover estimates for model calibration is laborious and very difficult to derive automatically. A number of studies, for instance, have explored the potential of delineat-ing individual tree crowns from aerial images and very high-resolution optical satellite imagery (Culvenor 2002; Erikson 2004; Jing et al. 2012; Song et al. 2010; Gessner et al. 2013), or based on lidar data (Riaño et al. 2004; Kato et al. 2009; Liu et al. 2013).

6. Conclusion

This study compared the effects of different explanatory and response variables on the accuracy of quantifying sub-pixel land-cover composition and continuous fields of tree cover for an area with distinct dry and wet seasons. In addition to model validation on a distant and a near test set, inter-annual variability of model predictions was analysed for an extended study area based on 11 years of MODIS data. The study is spatially limited to the lower Mekong Basin in Southeast Asia. However, certain conclusions can be drawn for other areas with similar seasonal patterns.

- Prediction accuracy is closely linked to cloud-cover distribution and deteriorates rapidly with the beginning of the wet season. The highest accuracy is achieved for the dry-season months where cloud cover is the lowest and the respective compo-sites are characterized by a minimum level of noise.
- The inclusion of temporal information obtained from the entire year, such as phenological metrics or EVI time series, did not facilitate any improvements in model accuracy when compared to the exclusive use of information during the dry

season. However, comparative studies showed that the contribution of multi-temporal information strongly depends on temporal resolution of the data used, climatic conditions, and the respective land-cover distribution within the study area.

- The exclusive use of the dry-season composites resulted in lower inter-annual variation compared to the shorter monthly composites for the same period or when phenological related information was included in the model. This finding is particularly relevant for long-term change analyses. By minimizing inter-annual fluctuations related to noise and external effects, more subtitle differences in, for instance, tree-cover estimates can be detected from year to year.
- The quality of the training and test fractions has shown to have significant effects on the accuracy of prediction results. This highlights the fact that the quality of the descriptive variable may be the most sensitive and most significant factor for sub-pixel mapping exercises.

References

Adams, J. B., D. E. Sabol, V. Kapos, R. A. Filho, D. A. Roberts, M. O. Smith, and A. R. Gillespie. 1995. "Classification of Multispectral Images Based on Fractions of Endmembers: Application to Land-Cover Change in the Brazilian Amazon." *Remote Sensing of Environment* 52 (2): 137–154. doi:10.1016/0034-4257(94)00098-8.

Braswell, B. H., S. C. Hagen, S. E. Frolking, and W. A. Salas. 2003. "A Multivariable Approach for Mapping Sub-Pixel Land Cover Distributions Using MISR and MODIS: Application in the Brazilian Amazon Region." *Remote Sensing of Environment* 87 (2–3): 243–256.

Breiman, L. 2001. "Random Forests." *Machine Learning* 45 (1): 5–32. doi:10.1023/A:1010933404324.

Breiman, L., J. Friedman, R. Olshen, and C. J. Stone. 1984. *Classification and Regression Trees.* New York, NY: Chapman and Hall.

Carpenter, G. A., S. Gopal, S. Macomber, S. Martens, and C. E. Woodcock. 1999. "A Neural Network Method for Mixture Estimation for Vegetation Mapping." *Remote Sensing of Environment* 70 (2): 138–152. doi:10.1016/S0034-4257(99)00027-9.

Cortes, C., and V. Vapnik. 1995. "Support-Vector Networks." *Machine Learning* 20 (3): 273–297. doi:10.1007/BF00994018.

Culvenor, D. S. 2002. "TIDA: An Algorithm for the Delineation of Tree Crowns in High Spatial Resolution Remotely Sensed Imagery." *Computers & Geosciences* 28 (1): 33–44. doi:10.1016/S0098-3004(00)00110-2.

DeFries, R. S., C. B. Field, I. Fung, C. O. Justice, S. Los, P. A. Matson, E. Matthews, H. A. Mooney, C. S. Potter, K. Prentice, P. J. Sellers, J. R. G. Townshend, C. J. Tucker, S. L. Ustin, and P. M. Vitousek. 1995. "Mapping the Land Surface for Global Atmosphere-Biosphere Models: Toward Continuous Distributions of Vegetation's Functional Properties." *Journal of Geophysical Research: Atmospheres* 100 (D10): 20867–20882. doi:10.1029/95JD01536.

DeFries, R. S., M. C. Hansen, M. Steininger, R. Dubayah, R. Sohlberg, and J. R. G. Townshend. 1997. "Subpixel Forest Cover in Central Africa from Multisensor, Multitemporal Data." *Remote Sensing of Environment* 60 (3): 228–246. doi:10.1016/S0034-4257(96)00119-8.

DeFries, R. S., M. C. Hansen, and J. R. G. Townshend. 1995. "Global Discrimination of Land Cover Types From Metrics Derived from AVHRR Pathfinder Data." *Remote Sensing of Environment* 54: 209–222. doi:10.1016/0034-4257(95)00142-5.

DeFries, R. S., M. C. Hansen, and J. R. G. Townshend. 2000. "Global Continuous Fields of Vegetation Characteristics: A Linear Mixture Model Applied to Multi-Year 8 km AVHRR Data." *International Journal of Remote Sensing* 21 (6–7): 1389–1414. doi:10.1080/014311600210236.

DeFries, R. S., M. C. Hansen, J. R. G. Townshend, and R. Sohlberg. 1998. "Global Land Cover Classifications at 8 km Spatial Resolution: The Use of Training Data Derived from Landsat Imagery in Decision Tree Classifiers." *International Journal of Remote Sensing* 19 (16): 3141–3168. doi:10.1080/014311698214235.

DeFries, R. S., J. R. G. Townshend, and M. C. Hansen. 1999. "Continuous Fields of Vegetation Characteristics at the Global Scale at 1-km Resolution." *Journal of Geophysical Research* 104 (D14): 16911–16923. doi:10.1029/1999JD900057.

Erikson, M. 2004. "Species Classification of Individually Segmented Tree Crowns in High-Resolution Aerial Images Using Radiometric and Morphologic Image Measures." *Remote Sensing of Environment* 91 (3–4): 469–477.

Esch, T., V. Himmler, G. Schorcht, M. Thiel, T. Wehrmann, F. Bachofer, C. Conrad, M. Schmidt, and S. Dech. 2009. "Large-Area Assessment Of Impervious Surface Based on Integrated Analysis of Single-Date Landsat-7 Images and Geospatial Vector Data." *Remote Sensing of Environment* 113 (8): 1678–1690. doi:10.1016/j.rse.2009.03.012.

Fernandes, R., R. Fraser, R. Latifovic, J. Cihlar, J. Beaubien, and Y. Du. 2004. "Approaches to Fractional Land Cover and Continuous Field Mapping: A Comparative Assessment Over the BOREAS Study Region." *Remote Sensing of Environment* 89 (2): 234–251. doi:10.1016/j.rse.2002.06.006.

Foody, G. M., and D. P. Cox. 1994. "Sub-Pixel Land Cover Composition Estimation Using a Linear Mixture Model and Fuzzy Membership Functions." *International Journal of Remote Sensing* 15 (3): 619–631. doi:10.1080/01431169408954100.

Foody, G. M., R. M. Lucas, P. J. Curran, and M. Honzak. 1997. "Non-Linear Mixture Modelling Without End-Members Using an Artificial Neural Network." *International Journal of Remote Sensing* 18 (4): 937–953. doi:10.1080/014311697218845.

Friedl, M. A., C. Woodcock, S. Gopal, D. Muchoney, A. H. Strahler, and C. Barker-Schaaf. 2000. "A Note On Procedures Used For Accuracy Assessment in Land Cover Maps Derived from AVHRR Data." *International Journal of Remote Sensing* 21 (5): 1073–1077. doi:10.1080/014311600210434.

Gessner, U., M. Machwitz, C. Conrad, and S. Dech. 2013. "Estimating the Fractional Cover of Growth Forms and Bare Surface In Savannas. A Multi-Resolution Approach Based on Regression Tree Ensembles." *Remote Sensing of Environment* 129: 90–102. doi:10.1016/j.rse.2012.10.026.

Guerschman, J. P., M. J. Hill, L. J. Renzullo, D. J. Barrett, A. S. Marks, and E. J. Botha. 2009. "Estimating Fractional Cover of Photosynthetic Vegetation, Non-Photosynthetic Vegetation and Bare Soil in the Australian Tropical Savanna Region Upscaling the EO-1 Hyperion and MODIS Sensors." *Remote Sensing of Environment* 113 (5): 928–945. doi:10.1016/j.rse.2009.01.006.

Hall, M., E. Frank, G. Holmes, B. Pfahringer, P. Reutemann, and I. H. Witten. 2009. "The WEKA Data Mining Software: An Update." *ACM SIGKDD Explorations Newsletter* 11 (1): 10–18. doi:10.1145/1656274.1656278.

Hansen, M. C., R. S. DeFries, J. R. G. Townshend, R. Sohlberg, C. Dimiceli, and M. Carroll. 2002. "Towards an Operational MODIS Continuous Field of Percent Tree Cover Algorithm: Examples Using AVHRR and MODIS Data." *Remote Sensing of Environment* 83 (1–2): 303–319.

Hansen, M. C., J. R. G. Townshend, R. S. DeFries, and M. Carroll. 2005. "Estimation of Tree Cover Using MODIS Data at Global, Continental and Regional/Local Scales." *International Journal of Remote Sensing* 26 (19): 4359–4380. doi:10.1080/01431160500113435.

Houghton, R. 1994. "The Worldwide Extent of Land-Use Change." *BioScience* 44 (5): 305–313. doi:10.2307/1312380.

Huete, A., K. Didan, T. Miura, E. Rodriguez, X. Gao, and L. Ferreira. 2002. "Overview of the Radiometric and Biophysical Performance of the MODIS Vegetation Indices." *Remote Sensing of Environment* 83 (1–2): 195–213.

Irish, R. R., J. L. Barker, S. N. Goward, and T. Arvidson. 2006. "Characterization of the Landsat-7 ETM+ Automated Cloud-Cover Assessment (ACCA) Algorithm." *Photogrammetric Engineering & Remote Sensing* 72 (10): 1179–1188. doi:10.14358/PERS.72.10.1179.

Jing, L., B. Hu, T. Noland, and J. Li. 2012. "An Individual Tree Crown Delineation Method Based on Multi-Scale Segmentation of Imagery." *Isprs Journal of Photogrammetry and Remote Sensing* 70: 88–98. doi:10.1016/j.isprsjprs.2012.04.003.

Kato, A., L. M. Moskal, P. Schiess, M. E. Swanson, D. Calhoun, and W. Stuetzle. 2009. "Capturing Tree Crown Formation Through Implicit Surface Reconstruction Using Airborne Lidar Data." *Remote Sensing of Environment* 113 (6): 1148–1162. doi:10.1016/j.rse.2009.02.010.

Kruse, F. A., A. B. Lefkoff, J. W. Boardman, K. B. Heidebrecht, A. T. Shapiro, P. J. Barloon, and A. F. H. Goetz. 1993. "The Spectral Image Processing System (SIPS) Interactive Visualization and

Analysis of Imaging Spectrometer Data." *Remote Sensing of Environment* 44: 145–163. doi:10.1016/0034-4257(93)90013-N.

Kuenzer, C., M. Bachmann, A. Mueller, L. Lieckfeld, and W. Wagner. 2008. "Partial Unmixing as a Tool for Single Surface Class Detection and Time Series Analysis." *International Journal of Remote Sensing* 29 (11): 3233–3255. doi:10.1080/01431160701469107.

Kumar, U., N. Kerle, T. V. Ramachandra, and K. Elleithy. 2008. "Constrained Linear Spectral Unmixing Technique for Regional Land Cover Mapping Using MODIS Data." *Innovations and Advanced Techniques in Systems, Computing Sciences and Software Engineering* 416–423. doi:10.1007/978-1-4020-8735-6_78.

Lambin, E. F., B. L. Turner, H. J. Geist, S. B. Agbola, A. Angelsen, J. W. Bruce, O. T. Coomes, R. Dirzo, G. Fischer, C. Folke, P. S. George, K. Homewood, J. Imbernon, R. Leemans, X. Li, E. F. Moran, M. Mortimore, P. S. Ramakrishnan, J. F. Richards, H. Skånes, W. Steffen, G. D. Stone, U. Svedin, T. A. Veldkamp, C. Vogel, and J. Xu. 2001. "The Causes of Land-Use and Land-Cover Change: Moving Beyond the Myths." *Global Environmental Change* 11 (4): 261–269. doi:10.1016/S0959-3780(01)00007-3.

Leinenkugel, P., T. Esch, and C. Kuenzer. 2011. "Settlement Detection and Impervious Surface Estimation in the Mekong Delta Using Optical and SAR Remote Sensing Data." *Remote Sensing of Environment* 115 (12): 3007–3019. doi:10.1016/j.rse.2011.06.004.

Leinenkugel, P., C. Kuenzer, and S. Dech. 2013. "Comparison and Enhancement of MODIS Cloud Mask Products for Southeast Asia." *International Journal of Remote Sensing* 34 (8): 2730–2748. doi:10.1080/01431161.2012.750037.

Leinenkugel, P., C. Kuenzer, N. Oppelt, and S. Dech. 2013. "Characterisation of Land Surface Phenology and Land Cover Based on Moderate Resolution Satellite Data in Cloud Prone Areas —A Novel Product for the Mekong Basin." *Remote Sensing of Environment* 136: 180–198. doi:10.1016/j.rse.2013.05.004.

Liu, W., K. C. Seto, E. Y. Wu, S. Gopal, and C. E. Woodcock. 2004. "ART-MMAP: A Neural Network Approach to Subpixel Classification." *IEEE Transactions on Geoscience and Remote Sensing* 42 (9): 1976–1983. doi:10.1109/TGRS.2004.831893.

Liu, J., J. Shen, R. Zhao, and S. Xu. 2013. "Extraction Of Individual Tree Crowns from Airborne LiDAR Data in Human Settlements." *Mathematical and Computer Modelling* 58 (3–4): 524–535.

Liu, W., and E. Y. Wu. 2005. "Comparison of Non-Linear Mixture Models: Sub-Pixel Classification." *Remote Sensing of Environment* 94 (2): 145–154. doi:10.1016/j.rse.2004.09.004.

Lu, D., E. Moran, and M. Batistella. 2003. "Linear Mixture Model Applied to Amazonian Vegetation Classification." *Remote Sensing of Environment* 87 (4): 456–469. doi:10.1016/j.rse.2002.06.001.

Lu, D., and Q. Weng. 2007. "A Survey of Image Classification Methods and Techniques for Improving Classification Performance." *International Journal of Remote Sensing* 28 (5): 823–870. doi:10.1080/01431160600746456.

Matthews, E. 1983. "Global Vegetation and Land Use: New High-Resolution Data Bases for Climate Studies." *Journal of Climate and Applied Meteorology* 22: 474–487. doi:10.1175/1520-0450(1983)0222.0.CO;2.

Nelson, R., and B. Holben. 1986. "Identifying Deforestation in Brazil Using Multiresolution Satellite Data." *International Journal of Remote Sensing* 7 (3): 429–448. doi:10.1080/01431168608954696.

Quinlan, J. R. 1992. "Learning with Continuous Classes." AI '92: Proceedings of the 5th Australian Joint Conference on Artificial Intelligence, Hobart, 343–348.

Quinlan, J. R. 1993. *C4.5: Programs for Machine Learning*. San Francisco, CA: Morgan Kaufmann.

Riaño, D., E. Chuvieco, S. Condés, J. González-Matesanz, and S. L. Ustin. 2004. "Generation of Crown Bulk Density for Pinus Sylvestris L. from Lidar." *Remote Sensing of Environment* 92 (3): 345–352. doi:10.1016/j.rse.2003.12.014.

Richter, R. 1997. "Correction of Atmospheric and Topographic Effects for High Spatial Resolution Satellite Imagery." *International Journal of Remote Sensing* 18 (5): 1099–1111. doi:10.1080/014311697218593.

Scanlon, T. M., J. D. Albertson, K. K. Caylor, and C. A. Williams. 2002. "Determining Land Surface Fractional Cover from NDVI and Rainfall Time Series for a Savanna Ecosystem." *Remote Sensing of Environment* 82 (2–3): 376–388.

Schaaf, C. B., F. Gao, A. H. Strahler, W. Lucht, X. Li, T. Tsang, N. C. Strugnell, X. Zhang, Y. Jin, J.-P. Muller, P. Lewis, M. Barnsley, P. Hobson, M. Disney, G. Roberts, M. Dunderdale, C. Doll, R. P. D'entremont, B. Hu, S. Liang, J. L. Privette, and D. Roy. 2002. "First Operational BRDF, Albedo Nadir Reflectance Products from MODIS." *Remote Sensing of Environment* 83: 135–148.

Schwarz, M., and N. E. Zimmermann. 2005. "A New GLM-Based Method for Mapping Tree Cover Continuous Fields Using Regional MODIS Reflectance Data." *Remote Sensing of Environment* 95 (4): 428–443. doi:10.1016/j.rse.2004.12.010.

Shao, Y., and R. S. Lunetta. 2011. "Sub-Pixel Mapping Of Tree Canopy, Impervious Surfaces, and Cropland in the Laurentian Great Lakes Basin Using MODIS Time-Series Data." *IEEE Journal of Selected Topics in Applied Earth Observations and Remote Sensing* 4 (2): 336–347. doi:10.1109/JSTARS.2010.2062173.

Song, C., M. B. Dickinson, L. Su, S. Zhang, and D. Yaussey. 2010. "Estimating Average Tree Crown Size Using Spatial Information from Ikonos and QuickBird Images: Across-Sensor and Across-Site Comparisons." *Remote Sensing of Environment* 114 (5): 1099–1107. doi:10.1016/j.rse.2009.12.022.

Tottrup, C., M. S. Rasmussen, L. Eklundh, and P. Jönsson. 2007. "Mapping Fractional Forest Cover Across the Highlands of Mainland Southeast Asia Using MODIS Data and Regression Tree Modelling." *International Journal of Remote Sensing* 28 (1): 23–46. doi:10.1080/01431160600784218.

Townshend, J. R. G., C. Huang, S. N. V. Kalluri, R. S. Defries, S. Liang, and K. Yang. 2000. "Beware of Per-Pixel Characterization of Land Cover." *International Journal of Remote Sensing* 21 (4): 839–843. doi:10.1080/014311600210641.

Vancutsem, C., J.-F. Pekel, P. Bogaert, and P. Defourny. 2007. "Mean Compositing, an Alternative Strategy for Producing Temporal Syntheses. Concepts and Performance Assessment for SPOT VEGETATION Time Series." *International Journal of Remote Sensing* 28: 5123–5141.

Vermote, E. F., N. Z. El Saleous, and C. O. Justice. 2002. "Atmospheric Correction of MODIS Data in the Visible To Middle Infrared: First Results." *Remote Sensing of Environment* 83 (1–2): 97–111.

Wang, Y., and I. H. Witten. 1996. *Induction of Model Trees for Predicting Continuous Classes.* Working paper 96/23. Hamilton: University of Waikato.

White, M. A., J. D. Shaw, and R. D. Ramsey. 2005. "Accuracy Assessment of the Vegetation Continuous Field Tree Cover Product Using 3954 Ground Plots in the South-Western USA." *International Journal of Remote Sensing* 26 (12): 2699–2704. doi:10.1080/01431160500080626.

Wilson, M., and A. Henderson-Sellers. 1985. "A Global Archive of Land Cover and Soils Data for Use in General Circulation Climate Models." *Journal of Climatology* 5: 119–143. doi:10.1002/joc.3370050202.

Witten, I. H., F. Eibe, and M. A. Hall. 2011. *Data Mining: Practical Machine Learning Tools and Techniques.* Burlington: Elsevier.

Xiao, X., P. Dorovskoy, C. Biradar, and E. Bridge. 2011. "A Library of Georeferenced Photos from the Field." *EOS, Transactions American Geophysical Union* 92 (49): 453–454. doi:10.1029/2011EO490002.

Recent climate variability and its impact on precipitation, temperature, and vegetation dynamics in the Lancang River headwater area of China

Chong Huang[a], Yafei Li[b], Gaohuan Liu[a], Hailong Zhang[c], and Qingsheng Liu[a]

[a]State Key Laboratory of Resources and Environmental Information System, Institute of Geographic Sciences and Natural Resources Research, CAS, Beijing, China; [b]College of Air Traffic Management, Civil Aviation University of China, Tianjin, China; [c]State Key Laboratory of Remote Sensing Science, Institute of Remote Sensing and Digital Earth, CAS, Beijing, China

The alpine ecosystem is one of the most fragile ecosystems threatened by global climate change. The impact of climate variability on the vegetation dynamics of alpine ecosystems has become important in global change studies. In this study, spatially explicit gridded data, including the Moderate Resolution Imaging Spectroradiometer (MODIS) land-surface temperature (LST) product (MOD11A1/A2), the Tropical Rainfall Measuring Mission (TRMM) rainfall product (3B43), and MODIS net primary productivity (NPP) product (MOD17A3), together with meteorological observation data, were used to explore the spatio-temporal pattern of climate variability and its impact on vegetation dynamics from 2000 to 2012 in the Lancang River headwater area. We found that the variation patterns of LST, precipitation, and NPP in the study area showed remarkable spatial differences. From the northwest to the southeast the spatial variation of average annual LST exhibited a decreasing–increasing–decreasing–increasing pattern. At the same time, most of the study area exhibited an increasing LST during the growing season. The annual precipitation increased in the semi-arid northern part, whereas it decreased in the semi-humid southern part. The precipitation variability during the growing season has a pattern similar to the annual precipitation variability. Although the majority of the regions have seen an NPP increase from 2000 to 2012, the responses of the vegetation to the varied climate factors were spatially heterogeneous. The alpine–subalpine meadows in the high-altitude areas were more sensitive to climate variability in the growing season. It is argued that satellite remote-sensing products have great potential in investigating the impact of climate variability on vegetation dynamics at the finer scale, especially for the Lancang River headwater area with complex surface heterogeneity.

1. Introduction

Many studies have indicated a strong interaction between ecosystem changes and climate variability (Cao and Woodward 1998; Pielke et al. 1998). Vegetation in terrestrial ecosystems is considered an intermediate link among the pedosphere, atmosphere, and hydrosphere of the Earth system, and variation in vegetation coverage has exhibited the most sensitive response to climate variability (Zavaleta et al. 2003). Therefore, vegetation dynamics and their relationship with climate variability have become a hotspot in global change studies (Rees et al. 2001). A significant increase in temperature in the northern hemisphere, especially at mid to high altitudes, has been reported since the 1980s (IPCC 2001). Such an increasing temperature has resulted in a marked enhancement in terrestrial photosynthetic activity and biomass (Hansen et al. 1996; Tucker et al. 2001). In China, the connection between vegetation and global warming has been well-studied (Fang et al. 2001; Piao et al. 2004). Owing to the extension of the growing season of vegetation caused by global warming, the vegetation activity in most areas of China has also increased (Fang et al. 2003). Compared with other areas at the same latitude, the Tibetan Plateau has a simpler vegetation structure owing to its unique plateau environment and is generally regarded as an extremely fragile ecosystem (Zhong et al. 2010). Additionally, the impact of human activities on the Plateau is weak, making this region an ideal study area and a natural laboratory for exploring the response of vegetation under climate variability in natural conditions (Zhou et al. 2007).

Remote sensing provides an effective tool for monitoring the land-surface parameters of large, complicated ecosystems. Among all the remote-sensing parameters, net primary productivity (NPP), the normalized difference vegetation index (NDVI), and the enhanced vegetation index (EVI) are the most commonly used to evaluate vegetation growing status (Schultz and Halpert 1993; Goetz 1997; Ji and Peters 2005; Leinenkugel et al. 2013). In recent years, satellite remote-sensing products have been frequently used to establish relationships between vegetation indicators and climate parameters for the Tibetan Plateau (Zhang et al. 2007; Gao et al. 2009; Zhong et al. 2010). However, in these studies almost all climatic data were obtained from meteorological stations and the vegetation indices were retrieved only at the corresponding sites. Therefore, the observation data from a few stations cannot fully represent the characteristics of climate and vegetation changes across a large area. Although overall climate variability and vegetation response at a relatively large scale have been studied intensively, information about regional ecosystem responses at a finer scale is lacking (Zhang et al. 2011).

We have chosen to study the spatio-temporal variability of regional climate and its impacts on the ecosystems in the Lancang River headwater area of China during the period of 2000–2012. The Lancang–Mekong River's headwaters are located in the interior of the Tibetan Plateau. Under global and regional change, the fragile ecosystems in the headwater area exhibited remarkable variations in the past several decades (Gao et al. 2010). According to the hydrological records in the past decade, the spring monthly average flow, which is mainly supplied by snowmelt, increased by 10% from 2000 to 2010, whereas the summer flood peak amplitude decreased by nearly 70%. As an important ecological barrier, the headwater area plays an important role in sustaining the healthy ecosystems in the whole Lancang–Mekong basin. Analysis of climate variability and its impact on the vegetation dynamics in the headwater area are fundamental for exploring the long-term effects of climate variability on the large spatio-temporal scale, and will provide new insights into the sustainable development of the Lancang–Mekong basin. In this study, multi-sourced remote-sensing data products from 2000 to 2012, combined with meteorological observation data, were combined to analyse the spatio-temporal patterns in temperature, precipitation, and NPP, and to explore the heterogeneous responses of vegetation to temperature and precipitation variations in the Lancang River headwater area.

2. Study area

The Lancang–Mekong River originates in the Qinghai–Tibet Plateau of China, flows across China, Myanmar, Laos, Thailand, and Cambodia, and meets the ocean in Vietnam. Its headwater region (Figure 1(*a*)), draining an area of 61,600 km^2, is located in the interior Qinghai–Tibet Plateau at an average altitude of 4000–5000 m, even reaching

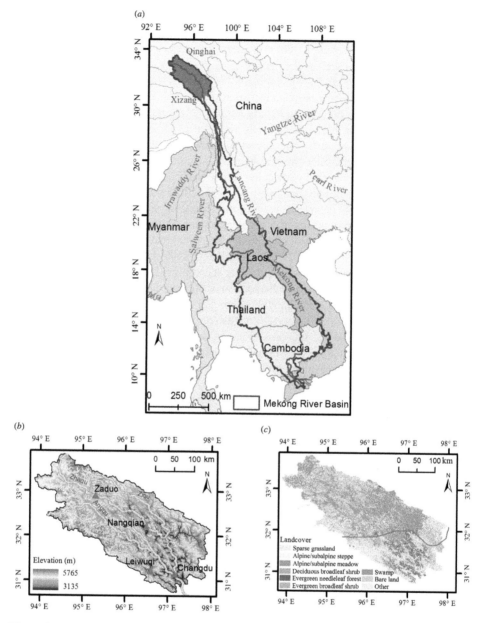

Figure 1. Lancang River headwater area. (*a*) Location of the study area in the Lancang–Mekong Basin. (*b*) DEM map of the study area. (*c*) Land-cover map of the study area (2010), the red line separates the different temperature zones, with the north part being the subarctic zone and the south part the temperate zone.

5500–6000 m in mountainous areas (Figure 1(*b*)). The average annual temperature and precipitation range from −3°C to 3°C and from 400 to 800 mm, respectively, and both gradually increase from the northwest to the southeast with an obvious vertical variation. Part of the source area in Qinghai Province has a cold semi-arid climate, whereas the Xizang part has a warmer and semi-humid climate. Accordingly, the vegetation types from the northwest to the southeast are alpine swamp, alpine–subalpine meadow, alpine–subalpine grassland, deciduous broadleaf shrub, and evergreen needleleaf forest (Figure 1(*c*)).

3. Materials and methods

3.1. Data collection

The meteorological data, collected at four China Meteorological Administration (CMA) weather stations (Zaduo, Nangqian, Leiwuqi, and Changdu) in the headwater area, were obtained from the Chinese National Meteorological Centre. The data included monthly mean surface air temperature and monthly total precipitation from 2000 to 2012. Figure 1(*b*) shows the locations of the four CMA stations over the study area. The stations were sparsely located at different altitudes, and represented the major land-cover types across the study area.

The satellite land-surface temperature data used in this study included MOD11A1, the daily level 3 land-surface temperature (LST) product at 1 km spatial resolution and MOD11A2, the averaged LSTs of the MOD11A1 product over eight days. All the images (from January 2000 to December 2012) were downloaded from the Land Processes Distributed Active Archive Center (https://lpdaac.usgs.gov/lpdaac/products/modis_products_table).

The Tropical Rainfall Measuring Mission (TRMM) Multisatellite Precipitation Analysis (TMPA) 3B43 data were applied to precipitation variability analysis in this study. Satellite-based precipitation estimates with high spatial and temporal resolutions and large areal coverage provide valuable information in areas scarcely gauged (Nicholson et al. 2003; Wolff et al. 2005; Su, Hong, and Lettenmaier 2008). The TRMM satellite was the first meteorological satellite specially designed for monitoring rainfall in the tropical and subtropical zones. TMPA is a calibration-based sequential scheme for combining precipitation estimates from multiple satellites, as well as gauge analyses where feasible, at fine scales of $0.25° \times 0.25°$ (Huffman et al. 2007). The monthly TRMM 3B43 data between 2000 and 2012 were downloaded from the National Aeronautics and Space Administration (NASA) website.

Annual NPP data used in this study (with a 1×1 km^2 spatial resolution, from 2000 to 2012) were processed by the Numerical Terradynamic Simulation Group (NTSG) (http://www.ntsg.umt.edu/project/mod17), based on the Earth Observing System (EOS) MODIS standard product MOD17A3. The MOD17 algorithm provides the first operational, near-real-time calculation of global NPP products from the EOS MODIS sensor (Zhao et al. 2005).

3.2. Data processing

3.2.1. Preprocessing the remote-sensing data products

(1) Reconstruction of time-series MODIS LST images

MODIS land products were preprocessed before they were downloaded. However, some cloud-contaminated LST pixels remain undetected by the NASA LST map production algorithms, which greatly degrade the LST quality and hamper its efficient applications (Wan 2008). The sequent climatic parameters were generated following reconstruction of time-series remote-sensing images during which the impact of low-quality pixels was reduced. The method for reconstructing time-series LST data was adopted from the methods used by Neteler (2010) and Ke, Song, and Ding (2012). Briefly, the accompanying quality assessment information was first applied to filter out invalid pixels. Next, an image-based histogram analysis was used to determine and remove pixels that show unusually low LST values. The residual invalid pixels were spatially interpolated with an algorithm based on regression analysis between surface temperature and digital elevation model (DEM) in each sliding window of the original image. More algorithm details is available in the literature (Neteler 2010; Ke, Song, and Ding 2012). Then the reconstructed eight day LST images were aggregated pixel-wise in geographical information system (GIS) software to obtain the monthly LST images. Finally, a simple average method was used in the current algorithm to produce the yearly and seasonal (from April to August) LST images.

(2) TRMM precipitation data re-sampling

The original monthly TRMM precipitation data have a spatial resolution of $0.25° \times 0.25°$. The data were re-sampled using a bilinear interpolation method to a resolution of 1×1 km^2 coherence to other remote-sensing products used in this study. Then the monthly data were aggregated by a summation method to derive the yearly images.

(3) Filling void pixels in the yearly MODIS NPP products

MODIS NPP products (MOD17A3) were used for analysing the inter-annual variation of NPP. The yearly MODIS NPP products use quality control (QC) to refer to the percentage of filled MOD15 LAI caused by cloud contamination during a growing season in a year (Zhao et al. 2005). By analysing the QC values of the 2000–2012 MOD17A3 products in the study area, we found that most areas with grassland cover have lower QC values, which means more reliable annual NPP. The invalid pixels that account for 2.4% of the total area were simply filled through the nearest interpolation algorithm.

3.2.2. Validation of remote-sensing data products based on ground meteorological data

The ground-based observation data were used as 'true value' to evaluate the information about the quality of MODIS LST and TRMM precipitation products. Based on the correlation coefficient and scatter point slope methods, the reliabilities of TRMM 3B43 data and MODIS LST data at a monthly timescale during 2000–2012 were validated using the ground measured data collected at the four meteorological stations in the Lancang River headwater area.

3.2.3. Linear trend analysis of NPP and climate variability

To analyse the spatio-temporal variability of climate factors and NPP during the study period, linear trend and slope of the time-series data were calculated using the least-squares method. A negative/positive regression coefficient indicates a decline/increase of the variable. The ratio of regression coefficient to the average variable value over the study period is defined as the mean increase/decrease rate of the annual variable. When the regression was applied pixel-by-pixel to the entire study area, the spatio-temporal variability of the climate factors and NPP was obtained.

4. Results

4.1. Evaluation of the utility of MODIS LST and TRMM precipitation data

Figures 2(a) and (b) represent the linear regression results of the ground measured air temperature (T_a)/precipitation (P) data from all four meteorological stations to the MODIS LST/TRMM precipitation (P_{TRMM}) data in the corresponding cell. From Figure 2(a), it is seen that the coefficient of determination R^2 for $P_{TRMM} - P$ is 0.8871, and the confidence interval is at the 95% confidence level. The results implied that TRMM data have an obvious linear correlation with ground-based precipitation data, although the TRMM precipitation is generally a bit smaller than the observation value. In Figure 2(b) the R^2 of LST $-$ T_a regression is 0.6249 at the 95% confidence level, which means that LST data have a linear correlation with ground-based air temperature. The regression operations were then implemented for each station (Figures 2(c) and (d) only considering the

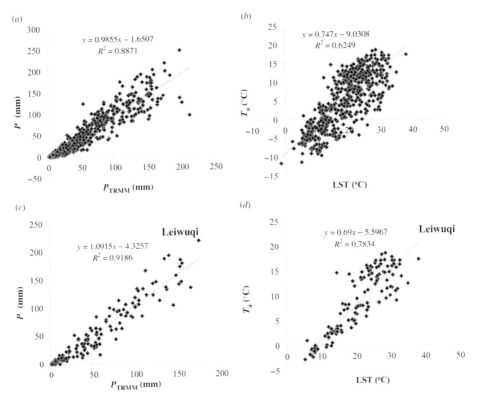

Figure 2. Linear regression of the ground-measured air temperature (T_a)/precipitation (P) (used as independent variable) to the MODIS LST/TRMM precipitation (P_{TRMM}) data in the corresponding cell (used as dependent variable). (a) $P_{TRMM} - P$, all four meteorological stations together; (b) LST $-$ T_a, all four meteorological stations together; (c) $P_{TRMM} - P$, for one station of Leiwuqi; and (d) LST $-$ T_a, for one station of Leiwuqi.

Leiwuqi station as an example), and the results were similar. In general, the linear correlation of $P_{\mathrm{TRMM}} - P$ is higher than that of $\mathrm{LST} - T_{\mathrm{a}}$.

4.2. Spatial and temporal patterns of temperature and precipitation variability in the headwater area

Based on the MODIS LST products and TRMM precipitation data of 2000–2012, we calculated the spatial patterns of the temperature and precipitation variations within the study area. As indicated in Figure 3(a), the variation rate of the average annual LST ($\mathrm{LST_A}$) is different across the whole study area. From the northwest to the southeast, the spatial variation of $\mathrm{LST_A}$ showed a decreasing-increasing-decreasing-increasing pattern. The regions with a downward $\mathrm{LST_A}$ accounted for approximately 55% of the entire study area. The rapid temperature decreases occurred in the northern areas with a sub-arctic climate, and the area proportion is 15.3%. The areas with a rapidly increasing $\mathrm{LST_A}$ amounted to 34.0% of the study area, mainly distributed in the southeast. In Figure 3(b) the LST in the growing season ($\mathrm{LST_{GS}}$) shows quite a different pattern compared with the $\mathrm{LST_A}$. Almost 81% of the study area exhibited LST increase in the growing season. The rapidly increasing $\mathrm{LST_{GS}}$ regions had a proportion of 30.6%, mainly distributed at the northwest, whereas the slightly increasing $\mathrm{LST_{GS}}$ areas accounted for 50.4%, mostly around the rapid-temperature-increase areas. At the same time, only 19% of the headwater area showed LST decrease in the growing season, mainly located at the northeast with a temperate climate.

As indicated in Figure 4(a), the variation of the annual TRMM precipitation between the northern and southern parts is opposite. The annual precipitation increased in the semi-arid northern part and decreased in the semi-humid southern part. However, the amplitude of precipitation variation is relatively small in most parts of the study area. The precipitation variability in the growing season had a similar pattern to the annual precipitation variability (Figure 4(b)). In some southern areas that had no vegetation cover, precipitation decrease in the growing season was remarkable.

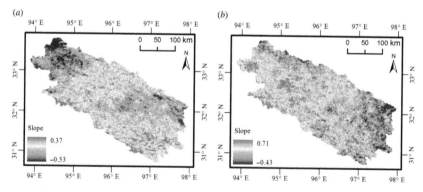

Figure 3.   LST variability pattern in the Lancang River headwater area between 2000 and 2012. (a) The variation slope of average annual LST ($\mathrm{LST_A}$); (b) the variation slope of LST in the growing season ($\mathrm{LST_{GS}}$). The red colour indicates a rapid increase and the blue colour represents a rapid decrease.

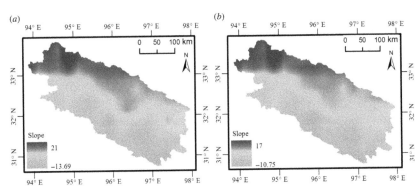

Figure 4.   TRMM precipitation variability pattern in the Lancang River headwater area between 2000 and 2012. (*a*) The variation slope of the annual TRMM precipitation; (*b*) the variation slope of TRMM precipitation in the growing season. The blue colour indicates a rapid increase and the green colour represents a rapid decrease.

4.3. *Annual NPP variation characteristics*

Through a statistical analysis of annual average NPP from 2000 to 2012, the overall temporal variation was obtained, which is shown in Figure 5(*a*). The highest NPP value (116.98 gC m^{-2} year^{-1}) in the study area occurred in 2010, and the lowest value (77.55 gC m^{-2} year^{-1}) occurred in 2003. In general, annual NPP in the study area gradually increased, although this variation was not stable. For example, NPP rapidly decreased during 2008 and 2011. As shown in Figure 5(*b*), the majority of the regions saw obvious NPP increase from 2000 to 2012. Areas with a slight increase, medium increase, and rapid increase in NPP accounted for 60.7%, 32.8%, and 5.9%, respectively. Regions with a decreasing NPP only accounted for 0.6% of the entire area. Slight NPP increase mainly occurred in the southern part with the land cover of temperate forest/shrub and the middle part with the cover of grassland, whereas the rapid NPP increase was mainly distributed in the northern part, with the cover of alpine-subalpine meadows.

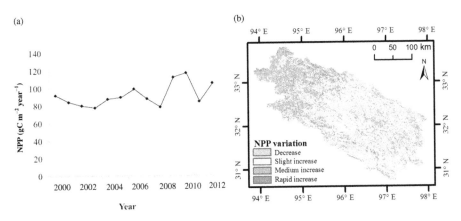

Figure 5.   Annual NPP change in the Lancang River headwater area from 2000 to 2012. (*a*) Temporal change of annual NPP; and (*b*) NPP variation pattern.

5. Discussion

5.1. Uncertainty analysis on the remote-sensing data products

The accuracy or error of MODIS land products and the TRMM precipitation products was evaluated by numerous studies, and the utilities of the products have been proved in many fields (Wan et al. 2004; Bookhagen and Burbank 2006; Fensholt et al. 2006; Zhao, Running, and Nemani 2006; Coll, Wan, and Galve 2009). For the majority of specific remote-sensing applications, analysing the data uncertainty and its effect on the results are of great importance (Leinenkugel, Kuenzer, and Dech 2013). The LST itself is more subject to variation than the air temperature measured at the ground meteorological stations. Specifically, the day/night LST algorithm is more sensitive to the cloud contaminations because the algorithm uses pairs of daytime and night-time MODIS observation data (Trigo et al. 2008; Wan 2008). It is not possible to make correct cloud masking in all cases, especially in the study areas with complex terrain and high surface heterogeneity. In this study, the eight day composite LST products of 2000–2012 were reconstructed. Validation based on four stations' observation data showed that the correlation coefficient of monthly average LST to T_a is around 0.9; however, some stations have relatively low correlation coefficients, such as Changdu (the coefficient of determination $R^2 = 0.5846$), where relatively dense shrubs existed. This meant that residual noise remained in the reconstructed monthly LST data. Furthermore, the error propagation in the aggregation operation might degrade the average annual LST data quality.

TMPA data integrate precipitation estimates from multiple satellites. However, the data quality to a large extent still depends on the quality and density of meteorological station observation data (Bookhagen and Burbank 2006). In general, the sparser the meteorological stations, the greater the difference in the data quality. In this study the TRMM data overestimated precipitation value in the northwest, where precipitation is relatively less and underestimated the precipitation value in the southeast with relatively abundant rainfall. However, the overall pattern of precipitation is reasonable according to the analysis of the meteorological observation data.

The yearly MODIS NPP products MOD17A3 were used as the indicator for analysing the vegetation dynamics in this study. Although some refinements and improvements were made to the MODIS primary vegetation productivity algorithm, the performance of the algorithm can be largely influenced by the uncertainties from inputs, such as land cover, FPAR/LAI (fraction of photosynthetically active radiation/leaf area index), the meteorological data, and the algorithm itself (Zhao et al. 2005). Through analysis of meaningful QC values for the yearly MODIS NPP products of the study area we found some invalid and low-quality pixels still existed. Simply filling the invalid pixels cannot improve the data quality itself. However, through visual inspection we found these invalid pixels were mainly distributed around the bare land or snow cover, so they might have little influence on the overall analysis of vegetation dynamics.

5.2. Spatial heterogeneity of climate variability in the headwater area

Although several studies have analysed the characteristics of climate variability at different timescales in the Tibetan Plateau (Gao et al. 2010; Zhong et al. 2010), little research has been conducted on climate variability across a relatively small region such as the Lancang River headwater area in the Qinghai Tibet Plateau. Regarding climate variability over the last decade in the Qinghai Tibet Plateau, almost all studies have concluded that this is a rapidly warming phase (Gao et al. 2009). Most of the research results depended on the analysis of time-series ground-based meteorological observation data. However, when considering the climate variability at a finer spatio-temporal scale, such analysis likely overlooked the spatial heterogeneity in climate variability. For example, all four meteorology stations in the study area exhibited increasing air temperature during the period 2000–2012. The coincidental results might imply that the whole region has experienced a temperature increasing phase. However, based on the spatially explicit remote-sensing LST data we found the temperature varied much across the landscape. From the northwest to the southeast, the study area can be divided into four subregions with different temperature variation tendencies. Since the Lancang River headwater area is only a small part of the Tibet Plateau, spatial heterogeneity in temperature can be mostly attributed to local geographical features such as the elevation, slope, aspect, and so on, or to the occasionality in meteorological conditions.

Compared with the generally accepted temperature increase, different researchers have obtained different or even contradictory results for the precipitation variation trend in the Tibetan Plateau (Zhang et al. 2011). Owing to the high spatial variability in precipitation, the observed data at meteorological stations can only represent the precipitation in a limited area, which might cause some discrepancy. According to this study, the precipitation variability showed a pattern of decrease in the northern part and increase in the southern part. Such spatial difference in precipitation may be associated with the temperature variability during the period. In the northern part, the decreasing temperature may cause more snowfall, whereas the rising temperature in the south may result in high evaporation and water vapour loss.

5.3. Response of NPP to climate variability in the headwater area

Studies have been carried out on the relationship between climate variability and NPP change in the Qinghai Tibet Plateau (Zhou et al. 2007; Zhong et al. 2010; Gao et al. 2013). It should be noted that almost all studies have compared NPP with climate variability based on data from scattered observation stations. Strictly speaking, the derived response of vegetation to climate variability can only be applied at the sites of these observation stations. When scaled up to an entire region, the conclusions may be uncertain or problematic. Our study indicates that owing to the spatial heterogeneity of temperature and precipitation, the response of NPP to climate variability exhibited clear differences across the Lancang River headwater area. Comparing the NPP and climate variability at the decade scale, we found that the average annual LST had a weak relationship with the vegetation NPP changes in the study area, whereas temperature variability in the growing season had a strong relationship with the alpine meadows NPP changes in the northwest. This may be attributed to the vegetation phenology in the study area. Under the control of Plateau climate, the alpine meadows start to grow in April as the temperature rises. During the period from September to March in the following year, plant photosynthesis and respiration are very weak due to the low temperature. The accumulation of NPP mainly occurs in the growing season from April to August. Thus, it is not difficult to understand that temperature variability in the growing season has more effect on NPP than the average annual temperature variation. This finding corroborates previously stated arguments that temperature rise has a strong influence on arctic, high-altitude regions (Hansen et al. 1996; IPCC 2001; Tucker et al. 2001). The effects of precipitation variability on vegetation NPP are quite complex. Annual precipitation had more effect on the vegetation NPP in the southeast, whereas precipitation in the growing season had a strong relationship with the NPP changes in the northwest. Considering the relatively slight variability of the precipitation during the short study period, the relationship between NPP and precipitation variability needs to be further studied.

6. Conclusions

Although overall climate variability and vegetation response at a relatively large scale have been studied intensively, information about regional ecosystem responses at a finer scale is lacking. Spatial heterogeneity is critical for analysing climate variability in a relatively small region. Remote-sensing-based monitoring in this study revealed the spatial heterogeneities both in the climate variability and in the responses of vegetation dynamics in the Lancang River headwater area from 2000 to 2012. From the northwest to the southeast, the spatial variation of average annual LST showed a decreasing–increasing–decreasing–increasing pattern. At the same time, most of the study area exhibited an increasing LST in the growing season. The variation of the annual TRMM precipitation is opposite between the northern and southern parts. The precipitation variability in the growing season had a pattern similar to the annual precipitation variability. Although the majority of the regions have seen an obvious NPP increase from 2000 to 2012, the responses of the vegetation to the varied climate factors show remarkable spatial heterogeneity. The alpine-subalpine meadows in the high-altitude areas were more sensitive to climate variability in the growing season at a shorter temporal scale, whereas annual precipitation had a greater impact on the NPP variation of the temperate shrub-forest.

One of the shortcomings in this article that must be noted is the relatively short period of remotely sensed data used, which is only from 2000 to 2012 during the MODIS period of record. Thus, some results have no strict interpretations of statistical significance. However, the research focused on climate or vegetation variability and their spatial heterogeneity across the landscape. Thus, this short data record is still useful because it is difficult to use individual meteorological observation to understand a highly varied region, especially for the rugged mountainous Lancang River headwater area. We argued that spatial heterogeneity is very important to study the regional response to global change since mitigation and adaptation strategies for global change are often applied at the regional scale. Our preliminary research revealed the response of vegetation to climate variability at a finer scale, which can be used for targeted adaptation strategies for ecosystem protection in the Lancang River headwater area.

Acknowledgements

The authors would like to thank two anonymous reviewers for their valuable comments on the manuscript.

Funding

We gratefully acknowledge the financial support from the National Natural Science Foundation of China [grant no. 40901223], Key Project for the 'One-Three-Five' Strategic Planning in IGSNRR [grant no. 2012ZD007], Special Programme for National Science and Technology Basic Research [grant no. 2008FY110300], and the Key Research Programme of the Chinese Academy of Sciences [grant no. KZZD-EW-08].

References

Bookhagen, B., and D. W. Burbank. 2006. "Topography, Relief, and TRMM-Derived Rainfall Variations Along the Himalaya." *Geophysical Research Letters* 33: L08405. doi:10.1029/2006GL026037.

Cao, M., and F. I. Woodward. 1998. "Dynamic Responses of Terrestrial Ecosystem Carbon Cycling to Global Climate Change." *Nature* 393: 249–252. doi:10.1038/30460.

Coll, C., Z. Wan, and J. M. Galve. 2009. "Temperature-Based and Radiance-Based Validations of the V5 MODIS Land Surface Temperature Product." *Journal of Geophysical Research* 114: D20102. doi:10.1029/2009JD012038.

Fang, J. Y., A. P. Chen, C. H. Peng, S. Q. Zhao, and L. J. Ci. 2001. "Changes in Forest Biomass Carbon Storage in China between 1949 and 1998." *Science* 292: 2320–2322. doi:10.1126/science.1058629.

Fang, J. Y., S. L. Piao, C. B. Field, Y. D. Pan, Q. H. Guo, L. M. Zhou, C. H. Peng, and S. Tao. 2003. "Increasing Net Primary Production in China from 1982 to 1999." *Frontiers in Ecology and the Environment* 1: 293–297. doi:10.1890/1540-9295(2003)001[0294:INPPIC]2.0.CO;2.

Fensholt, R., I. Sandholt, M. S. Rasmussen, S. Stisen, and A. Diouf. 2006. "Evaluation of Satellite Based Primary Production Modelling in the Semi-Arid Sahel." *Remote Sensing of Environment* 105: 173–188. doi:10.1016/j.rse.2006.06.011.

Gao, Q. Z., Y. Li, Y. Wan, X. Qin, W. Jiangcun, and Y. Liu. 2009. "Dynamics of Alpine Grassland NPP and its Response to Climate Change in Northern Tibet." *Climatic Change* 97: 515–528. doi:10.1007/s10584-009-9617-z.

Gao, Q. Z., Y. F. Wan, H. M. Xu, Y. Li, W. Z. Jiangcun, and A. Borjigidai. 2010. "Alpine Grassland Degradation Index and its Response to Recent Climate Variability in Northern Tibet, China." *Quaternary International* 226: 143–150. doi:10.1016/j.quaint.2009.10.035.

Gao, Y. H., X. Zhou, Q. Wang, C. Z. Wang, Z. M. Zhan, L. F. Chen, J. X. Yan, and R. Qu. 2013. "Vegetation Net Primary Productivity and its Response to Climate Change During 2001–2008 in the Tibetan Plateau." *Science of the Total Environment* 444: 356–362. doi:10.1016/j.scitotenv.2012.12.014.

Goetz, S. J. 1997. "Multi-Sensor Analysis of NDVI, Surface Temperature and Biophysical Variables at a Mixed Grassland Site." *International Journal of Remote Sensing* 18: 71–94. doi:10.1080/014311697219286.

Hansen, J., R. Ruedy, M. Sato, and R. Reynolds. 1996. "Global Surface Air Temperature in 1995: Return to Pre-Pinatubo Level." *Geophysical Research Letters* 23: 1665–1668. doi:10.1029/96GL01040.

Huffman, G. J., D. T. Bolvin, E. J. Nelkin, D. B. Wolff, R. F. Adler, G. J. Gu, Y. Hong, K. P. Bowman, and E. F. Stocker. 2007. "The TRMM Multisatellite Precipitation Analysis (TMPA): Quasi-Global, Multiyear, Combined-Sensor Precipitation Estimates at Fine Scales." *Journal of Hydrometeorology* 8: 38–55. doi:10.1175/JHM560.1.

IPCC. 2001. *Climate Change 2001: The Scientific Basis*, edited by J. T. Houghton, Y. Ding, D. J. Griggs, N. Noguer, P. J. van der Linden, X. Dai, K. Maskell, and C. A. Johnson, 881. Cambridge: Cambridge University Press.

Ji, L., and A. J. Peters. 2005. "Lag and Seasonality Considerations in Evaluating AVHRR NDVI Response to Precipitation." *Photogrammetric Engineering and Remote Sensing* 71: 1053–1061. doi:10.14358/PERS.71.9.1053.

Ke, L. H., C. Song, and X. Ding. 2012. "Reconstructing Complete MODIS LST Based on Temperature Gradients in Northeastern Qinghai-Tibet Plateau." IEEE international geoscience and remote sensing symposium (IGARSS), Munich, July 22–29. doi:10.1109/IGARSS.2012.6350664.

Leinenkugel, P., C. Kuenzer, and S. Dech. 2013. "Comparison and Enhancement of MODIS Cloud Mask Products for Southeast Asia." *International Journal of Remote Sensing* 34: 2730–2748. doi:10.1080/01431161.2012.750037.

Leinenkugel, P., C. Kuenzer, N. Oppelt, and S. Dech. 2013. "Characterisation of Land Surface Phenology and Land Cover Based on Moderate Resolution Satellite Data in Cloud-Prone Areas – A Novel Product for the Mekong Basin." *Remote Sensing of Environment* 136: 180–198. doi:10.1016/j.rse.2013.05.004.

Neteler, M. 2010. "Estimating Daily Land Surface Temperatures in Mountainous Environments by Reconstructed MODIS LST Data." *Remote Sensing* 2: 333–351. doi:10.3390/rs1020333.

Nicholson, S. E., B. Some, J. Mccollum, E. Nelkin, D. Klotter, Y. Berte, B. M. Diallo, I. Gaye, G. Kpabeba, O. Ndiaye, J. N. Noukpozounkou, M. M. Tanu, A. Thiam, A. A. Toure, and A. K. Traore. 2003. "Validation of TRMM and Other Rainfall Estimates with a High-Density Gauge Dataset for West Africa. Part II: Validation of TRMM Rainfall Products." *Journal of Applied Meteorology* 42: 1355–1368. doi:10.1175/1520-0450(2003)042XXX1355:VOTAORXXX2.0. CO;2.

Piao, S. L., J. Y. Fang, W. Ji, Q. Guo, J. Ke, and S. Tao. 2004. "Variation in a Satellite-Based Vegetation Index in Relation to Climate in China." *Journal of Vegetation Science* 15: 219–226. doi:10.1658/1100-9233(2004)015[0219:VIASVI]2.0.CO;2.

Pielke, R. A., R. Avissar, M. Raupach, A. J. Dolman, X. Zeng, and A. S. Denning. 1998. "Interactions between the Atmosphere and Terrestrial Ecosystems: Influence on Weather and Climate." *Global Change Biology* 4: 461–475. doi:10.1046/j.1365-2486.1998.t01-1-00176.x.

Rees, M., R. Condit, M. Crawley, S. Pacala, and D. Tilman. 2001. "Long-Term Studies of Vegetation Dynamics." *Science* 293: 650–655. doi:10.1126/science.1062586.

Schultz, P. A., and M. S. Halpert. 1993. "Global Correlation of Temperature, NDVI, and Precipitation." *Advances in Space Research* 13: 277–280. doi:10.1016/0273-1177(93)90559-T.

Su, F., Y. Hong, and D. P. Lettenmaier. 2008. "Evaluation of TRMM Multisatellite Precipitation Analysis (TMPA) and its Utility in Hydrologic Prediction in the La Plata Basin." *Journal of Hydrometeorology* 9: 622–640. doi:10.1175/2007JHM944.1.

Trigo, I. F., I. T. Monteiro, F. Olesen, and E. Kabsch. 2008. "An Assessment of Remotely Sensed Land Surface Temperature." *Journal of Geophysical Research* 113: D17108. doi:10.1029/2008JD010035.

Tucker, C. J., D. A. Slayback, J. E. Pinzon, S. O. Los, R. B. Myneni, and M. G. Taylor. 2001. "Higher Northern Latitude Normalized Difference Vegetation Index and Growing Season Trends from 1982 to 1999." *International Journal of Biometeorology* 45: 184–190. doi:10.1007/s00484-001-0109-8.

Wan, Z., Y. Zhang, Q. Zhang, and Z. L. Li. 2004. "Quality Assessment and Validation of the MODIS Global Land Surface Temperature." *International Journal of Remote Sensing* 25: 1, 261–274. doi:10.1080/0143116031000116417.

Wan, Z. M. 2008. "New Refinements and Validation of the MODIS Land-Surface Temperature/ Emissivity Products." *Remote Sensing of Environment* 112: 59–74. doi:10.1016/j. rse.2006.06.026.

Wolff, D. B., D. A. Marks, E. Amitai, D. S. Silberstein, B. L. Fisher, A. Tokay, J. Wang, and J. L. Pippitt. 2005. "Ground Validation for the Tropical Rainfall Measuring Mission (TRMM)." *Journal of Atmospheric and Oceanic Technology* 22: 365–380. doi:10.1175/JTECH1700.1.

Zavaleta, E. S., M. R. Shaw, N. R. Chiariello, H. A. Mooney, and C. B. Field. 2003. "Additive Effects of Simulated Climate Changes, Elevated CO_2, and Nitrogen Deposition on Grassland Diversity." *Proceedings of the National Academy of Sciences of the United States of America* 100: 7650–7654. doi:10.1073/pnas.0932734100.

Zhang, J. H., F. M. Yao, L. Y. Zheng, and L. M. Yang. 2007. "Evaluation of Grassland Dynamics in the Northern-Tibet Plateau of China Using Remote Sensing and Climate Data." *Sensors* 7: 3312–3328. doi:10.3390/s7123312.

Zhang, S. F., D. Hua, X. J. Meng, and Y. Y. Zhang. 2011. "Climate Change and its Driving Effect on the Runoff in the 'Three-River Headwaters' Region." *Acta Geographica Sinica* 66: 13–24. (in Chinese.)

Zhao, M., F. A. Heinsch, R. R. Nemani, and S. W. Running. 2005. "Improvements of the MODIS Terrestrial Gross and Net Primary Production Global Data Set." *Remote Sensing of Environment* 95: 164–176. doi:10.1016/j.rse.2004.12.011.

Zhao, M., S. W. Running, and R. R. Nemani. 2006. "Sensitivity of Moderate Resolution Imaging Spectroradiometer (MODIS) Terrestrial Primary Production to the Accuracy of Meteorological Reanalyses." *Journal of Geophysical Research* 111: G01002. doi:10.1029/2004JG000004.

Zhong, L., Y. M. Ma, M. S. Salama, and Z. B. Su. 2010. "Assessment of Vegetation Dynamics and their Response to Variations in Precipitation and Temperature in the Tibetan Plateau." *Climatic Change* 103: 519–535. doi:10.1007/s10584-009-9787-8.

Zhou, D. W., G. Z. Fan, R. H. Huang, Z. F. Fang, Y. Q. Liu, and H. Q. Li. 2007. "Interannual Variability of the Normalized Difference Vegetation Index on the Tibetan Plateau and its Relationship with Climate Change." *Advances in Atmospheric Sciences* 24: 474–484. doi:10.1007/s00376-007-0474-2.

Drought impact on vegetation productivity in the Lower Mekong Basin

Binghua Zhang[a,b], Li Zhang[a], Huadong Guo[a], Patrick Leinenkugel[c], Yu Zhou[a], Li Li[d], and Qian Shen[a]

[a]Key Laboratory of Digital Earth Science, Institute of Remote Sensing and Digital Earth, Chinese Academy of Sciences, Beijing, China; [b]College of Resources and Environment, University of Chinese Academy of Sciences, Beijing, China; [c]German Remote Sensing Data Center, DFD, German Earth Observation Center, DLR, Oberpfaffenhofen, Wessling, Germany; [d]College of Information Science and Engineering, Shandong Agricultural University, Tai'an, Shandong Province, China

The Lower Mekong Basin (LMB) has a typical monsoon climate, with high temperatures and an uneven distribution of precipitation throughout the year. This climate, combined with the geographic position of the LMB, has led to an increase in the frequency of extreme weather events over last decade. However, few previous studies have used remote-sensing data to investigate the impact of such weather events, particularly severe droughts, on biological productivity in the LMB. To address this, we assessed the impact of drought on vegetation productivity in the LMB during 2000–2011 using MOD17 products. Several drought events were identified during this period. Of these, the most severe occurred during 2005 and 2010, although the 2005 drought was both more extensive and more intense. Net primary productivity (NPP) exhibited considerable variation during 2000–2011: the droughts in 2005 and 2010 reduced NPP by 14.7% and 8.4%, respectively. The impact of drought on NPP in 2005 was much greater than that in 2010, likely owing to the longer duration and larger deficit of precipitation in 2005 (which lasted from winter 2004 to spring 2005). Our results demonstrate that severe drought had a greater impact on NPP than mild drought, especially for forests, woodlands, and shrublands. Comparatively, little variation in NPP was found for croplands, even under drought conditions, which were attributed to the wide use of irrigation and the exploitation of water sources during drought periods. Moreover, multi-season croplands in Vietnam experienced only a small reduction in gross primary productivity (GPP) in 2005 compared to one-season croplands in Cambodia, which can be related to the shorter growing periods of the former impacted by droughts.

1. Introduction

Global climate change has received increasing public attention during the past few decades owing to the associated loss of human life and increasing economic costs (Karl and Easterling 1999). In particular, extreme climate events have increased in both frequency and magnitude and are projected to become more frequent and severe during the remainder of the twenty-first century (IPCC 2007). Under the influence of both the Pacific and Indian Oceans, countries in Asia (particularly coastal Southeast Asia) are facing an increased frequency of climate change-related risks owing to the concentration of population and economic activity in coastal areas, the low per capita income, and high reliance on agriculture and natural resources (Paw and Thia-Eng 1991; ADB 2009; Sovacool et al. 2012). In fact, it has been predicted that, under a business-as-usual scenario, countries in Southeast Asia (including Indonesia, the Philippines, Thailand, and Vietnam) could lose 6.7% of their GDP annually owing to climate change; this is more than twice the global average loss (ADB 2009).

Increases in the frequency and magnitude of extreme climatic events (e.g. flood, drought, and tropical cyclones) could potentially have significant impacts on vegetation ecosystems through declining crop yields, delays in agricultural planting schedules, increased forest losses, higher plant vulnerability, and insect infestation and diseases (Rosenzweig et al. 2001; IPCC 2007; ADB 2009). Of these issues, the problems caused by drought in particular cannot be neglected. Drought is a complex natural disaster that has not yet been defined precisely and universally (Wilhite 2000). Unlike floods, typhoons, and storms, droughts have complex formation mechanisms, develop slowly over several months, and tend to be more widespread geographically (Te 2007). Severe drought can disturb the photosynthetic function of plants and even destroy vulnerable individuals (Hanson and Weltzin 2000); such conditions can also alter the distribution, composition, and abundance of different vegetation ecosystems, increase the probability of pest and disease infestations in forests (Hanson and Weltzin 2000), and trigger forest fires (Westerling et al. 2006) and tree mortality. In this manner, drought can cause large reductions in vegetation productivity (Hanson and Weltzin 2000; Ciais et al. 2005; Asner and Alencar 2010; Zhao and Running 2010; Chen et al. 2012), crop yields (Pantuwan et al. 2002), and carbon sequestration capability (Scott et al. 2009). On the global scale, a recent study found that a reduction in net primary productivity (NPP) in the southern hemisphere was very closely related to a drying trend between 2000 and 2009 (Zhao and Running 2010). Similarly, other studies at continental or national scales have reported drought-induced reductions in vegetation productivity in North America (Pennington and Collins 2007; Kwon et al. 2008; Noormets et al. 2009; Zhang et al. 2010), the Amazon region (Lewis et al. 2011; Potter et al. 2011), Europe (Rambal et al. 2003; Ciais et al. 2005; Pereira et al. 2007; Gilgen and Buchmann 2009), inner Asia (Mohammat et al. 2012), and China (Saigusa et al. 2010; Zhang, Xiao, et al. 2012; Pei et al. 2013; Zhou et al. 2013).

The LMB is located in the tropical monsoon ecosystem and is affected greatly by the monsoon climate (Xue, Liu, and Ge 2011). About 90% of the precipitation in this region is concentrated during the wet periods; this uneven temporal distribution leads to extensive flooding during the wet season and water shortages during the dry season (Leinenkugel, Kuenzer, and Dech 2013). Accordingly, the LMB is very vulnerable to climate change. Extreme floods and droughts, which are considered to be two of the most significant consequences of climate change, have occurred frequently in the LMB during the last decade and are projected to become increasingly frequent in future (Eastham et al. 2008; MRC 2009). This predicted extreme weather is likely to have significant effects on the lives and livelihoods of people living in the region and on plant yield and economics (MRC 2003; Son et al. 2012; Kuenzer et al. 2013). Drought events occurred several times in the 1990s and have become a widespread concern for farmers in the LMB over recent decades (MRC 2005; Son et al. 2012). For example, drought during 1997–1998 and 2003–2005 caused water shortages and huge crop reductions, induced forest fires (IPCC 2007), and affected the livelihoods of local people (MRC 2010). In this context, investigation of the impact of drought on vegetation ecosystems in the LMB is crucial, particularly for the forests and croplands, which are primary sources of income for the majority of people living in this region.

Previously, a number of studies have focused on monitoring drought in the LMB using different indices. Zhou et al. (2011) analysed the spatial and temporal distributions of droughts during the period 1977–2010 in the Mekong River area based on a standardized precipitation index (SPI), whereas Buckley et al. (2007) analysed tree ring and climate data in conjunction with Palmer drought severity index (PDSI) data and suggested that El Niño might play an important role in controlling droughts in this region. Similarly, Naeimi et al. (2013) monitored spatial and temporal drought conditions in the LMB during 1991–2000 and 2007–2011 and demonstrated the high correlation between El Niño conditions and monthly soil moisture. Son et al. (2012) detected large drought-affected areas during the 2003–2006 dry seasons using MODIS normalized difference vegetation index (NDVI) and land-surface temperature (LST) data. However, the above studies focused primarily on monitoring drought using remote sensing or meteorological data, and few previous studies have investigated the impact of drought on vegetation productivity.

Productivity is a fundamental parameter in the study of ecology (Matsushita and Tamura 2002). In particular, it is integral to the understanding of carbon dynamics within the atmosphere–vegetation–soil continuum for different ecosystems and their responses to future climate change (Matsushita et al. 2004). NPP is a crucial ecological variable that represents the difference between the levels of carbon used in photosynthesis and respiration and quantifies the amount of carbon fixed by plants and accumulated as biomass (Field et al. 1998; Zhao and Running 2010). Similarly, gross primary productivity (GPP) is the total carbon gained by the system through net photosynthesis (Zeng et al. 2008). Most previous studies investigating Southeast Asia have focused on the analysis of productivity for local areas or specific vegetation types (Matsushita and Tamura 2002; Zhang, Ju, et al. 2012) or on the potential forces driving NPP variations. Studies of plant-related productivity focusing on climatic factors have typically found a clear relationship between productivity and temperature (Hirata et al. 2008) and the strength of the monsoon (Fu and Wen 1999) across East Asia. Furthermore, it has been shown that net ecosystem productivity is highly related to climate change associated with El Niño–Southern Oscillation (ENSO) events in tropical Asia (Patra et al. 2005), and large NPP anomalies in Southeast Asia have been associated (at least in part) with large reductions in photo-synthetically active radiation (PAR) owing to forest fires (Kobayashi, Matsunaga, and Hoyano 2005). In addition, anthropogenic factors such as land-use change were identified as playing an important role in carbon storage in monsoon Asia during 1860–1990 (Tian et al. 2003).

Although previous studies have focused on monitoring droughts and on characterizing productivity variations in LMB, few have focused on the impacts of extreme drought on vegetation productivity using remote-sensing methods. To address this, the present study aims to detect droughts in LMB and identify their impacts on vegetation productivity during 2000–2011. We believe that this will improve an understanding of the impacts of extreme weather events on tropical and subtropical forest and farmland ecosystems in LMB.

2. Data and methods

2.1. Study area

The LMB is located in the downstream area of the Mekong River, which is the longest river in Southeast Asia. The LMB encompasses parts of five countries: Myanmar, Lao People's Democratic Republic (Lao PDR), Thailand, Cambodia, and Vietnam (Figure 1).

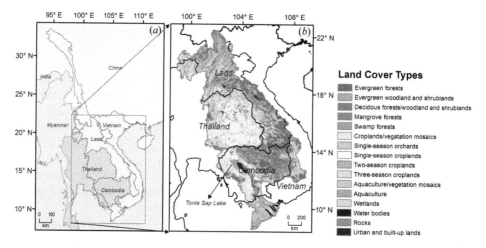

Figure 1. (a) Location of the Lower Mekong Basin and (b) spatial distribution of land-cover types. The boundary of the Mekong River Basin is adapted from Kuenzer et al. (2012). The land-cover product (Mekong LC2010) is adapted from Leinenkugel, Kuenzer, and Dech (2013).

The topography of the LMB is highly complex and can be divided into four parts: the Northern Highlands, Khorat Plateau, Tonle Sap Basin, and Mekong Delta (MRC 2010). Over 55 million people live in the LMB; of these, an estimated 75% depend on agriculture in combination with fishery, livestock, forestry, and other activities for their livelihoods (MRC 2010).

The LMB is characterized by a tropical monsoon climate influenced by the moist Southwest Monsoon and the dry Northeast Monsoon and experiences two distinct seasons: the wet season from June to October and the dry season from December to the following May (MRC 2010). This climatic pattern has a significant influence on plant growth in the region (Dudgeon 2000; Xue, Liu, and Ge 2011). As with its temporal distribution, the spatial distribution of precipitation is relatively heterogeneous, ranging from 1000 mm in northeast Thailand to 3000 mm in Cambodia (MRC 2003). Higher temperatures (average: 30–38°C) can be observed during March–April, with lower temperatures (average for the high-altitude region of Lao PDR: 15°C) during November–February (MRC 2003).

The Mekong River Basin is one of the most biodiverse regions globally and contains many ecoregions; accordingly, it is regarded as a crucial landscape of international biological importance (MRC 2010). Vegetation in the LMB varies depending on the distribution of precipitation and landforms; according to the Mekong LC2010 product, forests and croplands are two of the main vegetation types in the LMB (Leinenkugel et al. 2013). The spatial distribution of forests is controlled primarily by altitude and rainfall (MRC 2003), such that forests are found primarily in Lao PDR and Cambodia. Of the forests in the region, the greatest proportion is evergreen woodland and shrublands, and evergreen forests (16.4% and 20.3%, respectively), which are distributed primarily in Lao PDR and central Cambodia, followed by the deciduous forests/woodland and shrublands (8.9%) that are spread extensively throughout Cambodia. Unusual forest types, including swamp forests (1.5%) and mangrove forests (0.04%), are concentrated in small areas of the basin along Tonle Sap Lake and the Vietnam Mekong Delta, respectively. Croplands are distributed primarily within Thailand (single-season croplands, croplands/vegetation mosaics, and two-season croplands along the Mun River, which is the largest tributary of the Mekong River), Cambodia (single-season croplands and croplands/vegetation mosaics), and the Vietnam Mekong Delta (two-season croplands and three-season crop-lands). Croplands/vegetation mosaics constitute the highest areal proportion of the basin (29.2%), followed by single-season croplands (12.3%), two-season croplands (3.6%), and three-season croplands (0.8%). The most important crop type in the LMB is rice: rain-fed rice is distributed mainly in the central highlands of Vietnam, northeastern and parts of northern Thailand, and Cambodia, whereas fully or partially irrigated rice is grown in large parts of the Vietnam Mekong Delta (Nesbitt, Johnston, and Solieng 2003).

2.2. Data

2.2.1. Drought index

Drought indices have been used widely to detect drought intensity, duration, and spatial distribution. Drought characteristics and their impacts are very complex. Accordingly, various drought indices have been developed previously, including the Palmer drought severity index (PDSI) (Palmer 1965), drought area index (DAI) (Bhalme and Mooley 1980), Rainfall Anomaly Index (RAI) (Van Rooy 1965), and standardized precipitation index (SPI) (McKee, Doesken, and Kleist 1995). In the present study, we adopted the PDSI to detect drought conditions in the LMB during 2000–2011. We obtained the PDSI product at $0.5° \times 0.5°$ spatial resolution from the Numerical Terradynamic Simulation Group (ftp://ftp.ntsg.umt.edu/pub/MODIS/NTSG_Products).

PDSI is a good indicator of soil moisture fluctuations (Mika et al. 2005), where PDSI values of –1.0 to –1.9, –2.0 to –2.9, –3.0 to –3.9, and below –4.0 represent mild, moderate, severe, and extreme drought, respectively (Palmer 1965). PDSI was used originally to monitor drought in the USA and is now one of the most widely used drought-monitoring indices (Heim 2000). In particular, PDSI has been applied to detect drought in regions of the world including Europe (Domonkos, Szalai, and Zoboki 2001), Africa (Ntale and Gan 2003), and Brazil (Dos Santos and Pereira 1999). In contrast to many drought indices that are based solely on precipitation information, PDSI uses both precipitation and temperature as inputs. Previous studies have demonstrated that PDSI is correlated significantly with observed soil moisture data over large areas worldwide (Dai, Trenberth, and Qian 2004). A detailed description of PDSI can be found in Palmer (1965) and Alley (1984).

2.2.2. MODIS productivity products

Biological productivity represents productivity at different organic levels (e.g. individual, community, ecosystem, region, and biome) and indicates the overall health condition of ecological systems. Typically, GPP and NPP are two of the concepts representing biological productivity. In the present study, we obtained annual NPP information from the MOD17A3 NPP product and monthly GPP information from the MOD17A2 GPP product for the period 2000–2011 from the Numerical Terradynamic Simulation Group (http://www.ntsg.umt.edu). The current version (5.5) of the MODIS product offers improved data accuracy, achieved through the temporal infilling of cloud-contaminated pixels of MODIS MOD15 LAI/FPAR data products and the spatial interpolation of meteorological input data down to 1 km MODIS resolution. In addition, the modified Biome Parameter Look-Up Table (BPLUT), which is based on recent synthesized NPP data and observed GPP from flux tower measurements, has also been used to improve the MOD17 product's accuracy (Zhao et al. 2005; Zhao and Running 2010). The MOD17 product is the first regular, near-real-time satellite-driven dataset monitoring global vegetation productivity at 1 km resolution on a global scale (Zhao et al. 2005). The quality of the MOD17 product has been validated for different climatic regimes and vegetation types (Zhao, Running, and Nemani 2006; Turner et al. 2006), and this product has already been adopted in a wide variety of applications, including detecting drought impacts on vegetation productivity (Zhao and Running 2010; Shiba and Apan 2011; Zhang, Xiao, et al. 2012; Mu et al. 2013), calculating carbon-use efficiency (Kwon and Larsen 2013), and estimating wheat yield (Reeves, Zhao, and Running 2005). Although used frequently, the global MOD17 products have certain limitations. For example, the input data for NPP may be influenced by cloud (Zhao et al. 2005; Propastin et al. 2012), particularly because the average cloud cover in the LMB region reaches approximately 85–90% during the wet season (Leinenkugel, Kuenzer, and Dech 2013). Similarly, smoke from forest fires may also have some influence on the quality of the NPP product.

2.2.3. Climate data

We used monthly precipitation and temperature data from the Modern-Era Retrospective Analysis for Research and Applications (MERRA) dataset (Rienecker et al. 2011) for 2000–2011 to detect meteorological anomalies during the study period. These data are provided by the Global Modeling and Assimilation Office (GMAO) and cover the period from 1970 to the present with a spatial resolution of $0.500° \times 0.667°$. MERRA uses data derived from NASA's Earth Observation System satellites; this dataset has achieved significant improvements in reducing the uncertainty in precipitation and water vapour climatology through the representation of the atmospheric branch of the hydrological cycle in the reanalysis products (Rienecker et al. 2011).

2.2.4. Mekong LC2010 product

We used the Mekong LC2010 product (Leinenkugel et al. 2013) at a spatial resolution of 500 m to identify different vegetation types in LMB. This product is developed through a multi-step unsupervised classification approach based on MODIS spectral reflectance in bands 1–7 and information from an 11 year MODIS enhanced vegetation index (EVI) time series (2000–2011), including a series of phenological metrics for 2010. The accuracy of the Mekong LC2010 product was validated through comparison with numerous data, including Google Earth, Landsat TM, and auxiliary data products, with an overall accuracy of 74% (Leinenkugel et al. 2013). In fact, the quality of the Mekong LC2010 product has been shown to be superior to the 2009 GlobCover and 2009 MODIS land-cover (MCD12Q2) products, with overall accuracy approximately 20–25% greater than that of the other products (Leinenkugel et al. 2013).

2.3. Data analysis

2.3.1. Drought identification

We used PDSI data to characterize drought conditions across the basin for 2000–2011. In particular, we detected the impact of drought on productivity for mild drought areas with PDSI between −2 and −1 and severe drought areas with PDSI less than −2. The areal proportion affected by each of these levels of drought was calculated by dividing the area affected by drought by the total area of the entire LMB.

2.3.2. Drought impact on vegetation productivity

The MODIS NPP product was used to detect annual NPP anomalies for 2000–2011. We used the standardized anomalies index (SAI) to derive the spatial distribution of NPP anomalies and analysed the corresponding distribution of drought anomalies during the study period. In previous studies, SAI was applied effectively to detect anomalies in different variables (Giuffrida and Conte 1989; Peters et al. 2002). For example, Pei et al. (2013) applied SAI to assess the temporal dynamics of NPP anomalies in China and demonstrated the effectiveness of the index through its good correlation with the vegetation health index. In the present study, we defined SAI as follows:

$$SAI_{NPP} = \frac{NPP(i) - \overline{NPP}}{\sigma_{NPP}}. \tag{1}$$

Here, SAI_{NPP} represents the NPP anomalies, $NPP(i)$ is the annual NPP in the ith year, $\overline{NPP}$ is the value of mean NPP for period 2000–2011, and σ_{NPP} is the standard deviation of the 12 year NPP.

To investigate the impact of drought on NPP, we calculated the percentage of two levels of drought-affected areas ($-2 < PDSI < -1$ and $PDSI < -2$) and then analysed their correlations with the regional mean NPP for the study period. To provide a common resolution for NPP and GPP, we re-sampled the PDSI and climate data to a spatial resolution of 1 km. Then, we calculated the annual mean NPP for each of the dominant vegetation types and calculated relative changes in NPP (i.e. the difference between annual NPP and 12 year mean NPP, divided by the 12 year mean) under different levels of drought to analyse the responses of different vegetation types to different drought levels during 2000–2011.

To further explore the seasonal characteristics of the impact of drought on productivity during drought years, we calculated monthly mean GPP, mean temperature, and total precipitation in 2005 and 2010 and compared them with the 12 year means to identify meteorological anomalies and their impact on productivity in different seasons. We also selected the most drought-affected areas (Cambodia and Vietnam) to detect monthly GPP reductions for different vegetation types in 2005 and 2010.

3. Results and discussion

3.1. Droughts in the LMB over the last decade

During 2000–2011, the LMB experienced several drought events. Figure 2 illustrates the spatial distribution of PDSI, which characterizes the intensity and extent of drought for each year in the LMB during 2000–2011. These data demonstrate clearly that droughts occurred during 2003–2007 and in 2010. This observation is in accordance with the findings of Son et al. (2012), who used the NDVI and LST data to identify the 2003–2006 droughts (particularly that in 2005) that affected the majority of the LMB. Drought periods between 2003 and 2006 were also indicated by the MRC report (MRC 2005), whereas Naeimi et al. (2013) demonstrated recently that 2010 was also a drought year for the LMB. Thus, our observed results based on PDSI (Figure 2) are in accord with reported

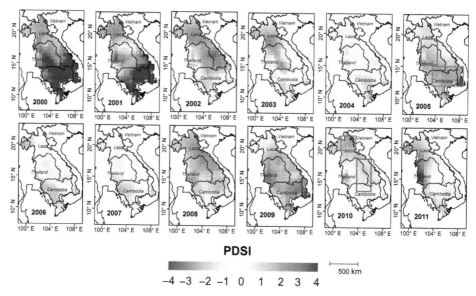

PDSI

−4 −3 −2 −1 0 1 2 3 4

500 km

Figure 2. Spatial distribution of PDSI over the period 2000–2011 over the Lower Mekong Basin. Blue and red boxes indicate examples of mild and severe drought, respectively.

drought events from other sources, demonstrating that PDSI can be considered an effective index for monitoring droughts in the LMB.

During 2003–2007 and 2010, mild drought hit 11.2%, 29.6%, 45.1%, 17.5%, 15.5%, and 38.3% of the LMB, respectively, whereas severe droughts typically affected only small portions of the LMB (2.9%, 6.3%, 2.4%, and 0.5% in 2003, 2004, 2006, and 2007, respectively) (Figure 6). Regionally, the drought in 2003 affected the Vietnam Mekong Delta and southwest Thailand, whereas that in 2004 affected primarily Cambodia and Vietnam. In 2006 and 2007, drought was observed in parts of Thailand and Lao PDR. In general, our results indicate that Cambodia has been affected by frequent droughts over the past decade, which is in accordance with the study of Son et al. (2012).

Relatively extensive and severe droughts occurred during 2005 and 2010 and affected almost the entire LMB. The regional mean values of PDSI during 2005 and 2010 were −2.1 and −1.1, respectively; these are the lowest values obtained throughout the study period (i.e. over 12 years) and are much lower than the long-term mean of 0.6 (Figure 3). In terms of both extent and intensity, the drought in 2005 was much more extensive and severe than that in 2010: the total area affected by drought in 2005 accounted for 98.1% of the LMB (52.9% for severe drought and 45.1% for mild drought), whereas that in 2010 accounted for only 58.7% of the LMB (20.2% for severe drought and 37.0% for mild drought). Furthermore, the regional mean PDSI in 2005 (−2.1) was much lower than that in 2010 (−1.1). The most severely affected region in 2005 was located in Cambodia, with an average PDSI of −2.5; this is much lower than the 12 year mean for Cambodia (0.8). Conversely, in 2010, the most severely affected region was Lao PDR, where the average PDSI was −1.9; this is considerably lower than the 12 year mean for Lao PDR (0.6).

Droughts in the LMB were induced primarily by high temperatures concurrent with shortages of precipitation (Figure 3). In the data for the LMB, precipitation shortages can be observed during 2003–2005 and 2009–2010, whereas relatively high temperatures can be observed for most of 2004–2006 and 2009–2010 (Figure 3). In 2005 and 2010, the

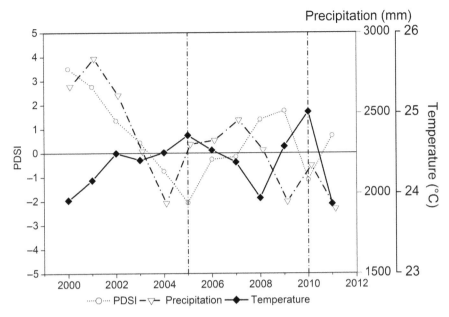

Figure 3. Inter-annual variations in regional mean PDSI, temperature, and precipitation over the period 2000–2011.

annual mean temperatures were 0.3°C and 0.6°C higher than the 2000–2011 mean, respectively. Moreover, annual total precipitation in 2010 was 5.9% lower than the 12 year mean. Although total precipitation in 2005 was close to the long-term mean, the shortage of water in the region can be traced back to October–December 2004, when the mean regional rainfall in LMB was only 57.2% of the 12 year mean and just 47% of the 1985–2000 normal rainfall (Te 2007).

3.2. *Drought impacts on annual NPP*

Figure 4 illustrates the annual NPP for the LMB during 2000–2011, which varied from nearly 142 g C m^{-2} year^{-1} in the Vietnam Mekong Delta to over 1521 g C m^{-2} year^{-1} in northern Lao PDR. Lower NPP was found along Tonle Sap Lake in Cambodia and the Vietnam Mekong Delta, where cropland types are dominant. In contrast, the dominant vegetation in Lao PDR is forests, which accounts for over 40% of the country's land area (Southavilay and Castrén 1999) and contributes to the high annul NPP in Lao PDR.

Figure 5 illustrates the annual NPP anomaly relative to the mean for the period 2000–2011. As can be seen from Figures 2 and 5, the spatial distribution of the PDSI is generally consistent with the distribution of the NPP anomalies in most years and areas. We found apparent negative anomalies in NPP for 2005 and 2010 for the entire region, with basin-averaged NPP values that were 14.7% and 8.4% less, respectively, than the 12 year mean. Moreover, the drought-affected areas were more extensive in these 2 years than in any other years studied (Figure 6). As we identified based on the PDSI (Figure 2), the NPP data indicate that the drought in 2005 was more severe and extensive than that in 2010. In particular, large areas of negative NPP anomalies were found in all of the countries studied in 2005 but were concentrated notably in Lao PDR and the east of Cambodia in 2010.

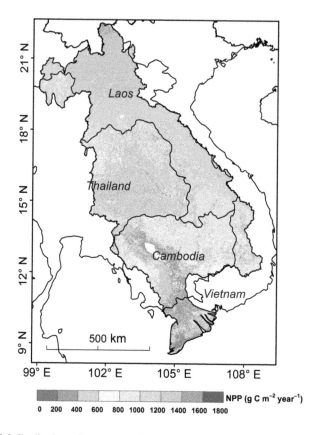

Figure 4. Spatial distribution of mean annual NPP over the period 2000–2011.

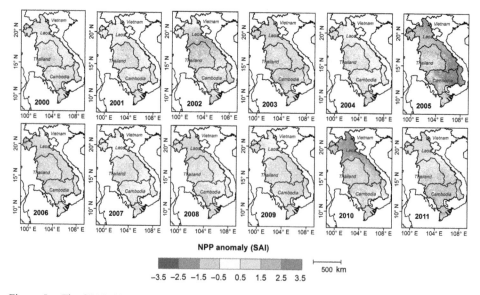

Figure 5. The 2000–2011 NPP anomalies relative to the mean (2000–2011).

98

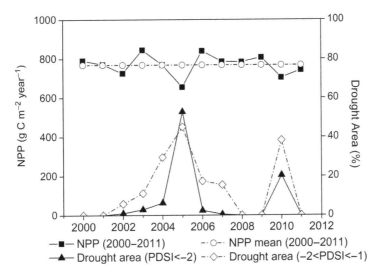

Figure 6. Inter-annual variations in regional mean NPP and areal percentage of drought-affected areas in the LMB from 2000 to 2011.

Compared with NPP in 2005 and 2010, the NPP data for other drought years (including 2003, 2004, 2006, and 2007) are not indicative of large reductions in NPP, which can be explained partly by the effects of different levels of drought. During 2003–2004 and 2006–2007, the percentages of areas experiencing severe drought (PDSI < −2) were relatively low (2.9%, 6.3%, 2.4%, and 0.5%, respectively) compared to those in 2005 and 2010; this could explain the relatively low NPP anomaly in these years. During the study period, the severe drought areas (PDSI < −2) exhibited a higher correlation with NPP ($R^2 = 0.55$) than did the mild drought areas (−2 < PDSI < −1; $R^2 = 0.31$) (Figure 6), indicating that severe drought had a much greater impact on regional NPP than mild drought. This has been demonstrated previously for China by Pei et al. (2013), who showed that a stronger correlation between droughts and NPP anomalies occurred during or after the time at which drought intensities reached their peak values.

In years for which no drought impact was identified by PDSI (e.g. 2002 and 2011), the negative NPP anomalies in the region (Figure 5) may be attributable to factors other than drought, such as the severe floods observed in the central LMB in 2002 and in the central and southern parts of the LMB in 2011 (MRC 2011). Several reasons for flooding-induced NPP reductions have been presented previously in the literature. For example, floods can have a significant impact on forests by inhibiting the growth of trees and suppressing respiration and on croplands through destruction of large areas of rice fields (Masumoto, Hai, and Shimizu 2008).

3.3. Drought impact on different vegetation ecosystems of the LMB

Figure 7 illustrates changes in NPP for the dominant vegetation types relative to the 12 year mean NPP during the study period. Drought (i.e. PDSI < −1) could not be observed in the LMB for 2000–2001, 2008–2009, or 2011; therefore, the relative NPP changes for these years were not considered. Large reductions in NPP were observed during 2005 and 2010 for all vegetation types, and severe droughts (PDSI < −2) had a greater impact on the NPP of different vegetation types (i.e. relative changes of 9.1–27.1% in 2005 and 4.0–

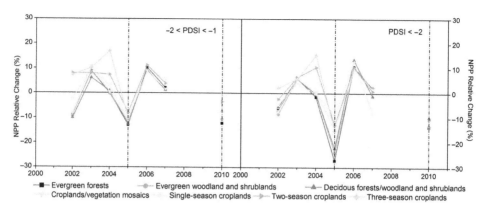

Figure 7. Relative NPP changes for different land-cover types for drought-affected areas where –2 < PDSI < –1 and PDSI < –2.0 during 2000–2011.

13.4% in 2010) than mild droughts (–2 < PDSI < –1; relative changes of 8.0–12.9% in 2005 and 0.2–12.1% in 2010).

Under the mild drought conditions of 2005, the relative change in NPP was small for forests, woodlands, and shrublands (12.9%, 11.8%, and 12.6% for evergreen forests, evergreen woodland and shrublands, and deciduous forests/woodland and shrublands, respectively), and croplands (11.0%, 10.6%, 8.0%, and 10.3% for croplands/vegetation mosaics, single-season croplands, two-season croplands, and three-season croplands, respectively). Conversely, under severe drought conditions, the impact of drought varied considerably for different vegetation types. For the four types of croplands, severe drought had a similar effect to that of mild drought: NPP declined by 14.5%, 9.2%, 12.8%, and 15.1% for croplands/vegetation mosaics, single-season croplands, two-season croplands, and three-season croplands, respectively, whereas the NPP reduction for forests, wood-lands, and shrublands were much larger (declines of 27.2%, 23.8%, and 22.1% for evergreen forests, evergreen woodland and shrublands, and deciduous forests/woodland and shrublands, respectively). This demonstrates that, while mild drought exerts relatively homogeneous effects regardless of vegetation type in LMB, the impact of severe droughts on forests, woodlands, and shrublands is much more pronounced than that on croplands. This lower impact for croplands may be attributable to the presence of irrigation systems that can mitigate the impact of droughts. Furthermore, severe droughts can be a potential factor in inducing forest fires (Taylor et al. 1999). In Vietnam, about 9000–12,000 ha of forests were burned during the 2002–2005 droughts (FAO 2007; ADB 2009), producing tree mortality over large areas (Allen et al. 2010) and reducing NPP. Moreover, smoke from forest fires reduces the amount of photosynthetically active radiation (PAR) available for plants, leading to further reductions in NPP (Kobayashi, Matsunaga, and Hoyano 2005). Our findings regarding the more significant impact of drought on forests than on croplands have also been observed in Amazonian forests: during the same years investigated in the present study (i.e. 2005 and 2010), Amazonian forests suffered severe drought that led to a large decline in productivity (Samanta et al. 2010; Lewis et al. 2011), which was particularly evident for closed broadleaf forests (Potter et al. 2011).

The relatively small reduction in NPP for croplands in the LMB can likely be ascribed to human factors. Rice is one of the most important agricultural products in the basin and accounts for more than 10 million ha of cultivated land (MRC 2010). In the Vietnam

Mekong Delta, about 60% of rice is irrigated (MRC 2010), and local people store water during the flood season and use it to irrigate rice during the drought period. In addition, improvements to irrigation systems (e.g. the addition of canal sluices to block sea water inflow) have also likely helped mitigate the effects of drought on irrigated rice (Sakamoto et al. 2009). Rain-fed rice is also important in the region investigated in the present study and is distributed primarily in Lao PDR, the central highlands of Vietnam, northeast and parts of northern Thailand, and Cambodia (Nesbitt, Johnston, and Solieng 2003). However, rain-fed rice is not typically affected seriously by drought, except for extreme drought, because water supplied through small ponds in and around the rain-fed fields can be used for irrigation during drought periods (Shimizu, Masumoto, and Pham 2006). Therefore, drought may not reduce crop production immediately for crop-intensive areas in LMB, where farmers can implement a series of adaptation strategies in case of insufficient precipitation. Consequently, NPP rates for croplands can be expected to be affected to a lesser degree than natural vegetation types (e.g. forests) by precipitation shortage and droughts.

3.4. Seasonal characterization of drought impact on productivity

To further investigate the seasonal characteristics of drought impact on vegetation productivity, we used monthly GPP data from the most severe drought years (2005 and 2010) to compare differences in seasonal reactions of vegetation during the 2 years. Figure 8 illustrates the variation in monthly GPP (along with temperature and precipitation) in the two severe drought years and previous years. Drought in 2005 was caused primarily by delayed monsoon rainfall and 2005 has been characterized as an El Niño year (Räsänen and Kummu 2012), which may contribute to the weather anomaly during this period. The water shortage in 2005 can be traced back to June 2004 and lasted until May 2005; precipitation was reduced by 19.9% during this period. In addition to this lower precipitation, higher temperatures occurred during the winter (January–February) of 2005; in particular, the monthly mean temperature for February 2005 was 2.4°C higher than the 12 year mean. The lack of precipitation associated with the 2010 drought can be traced

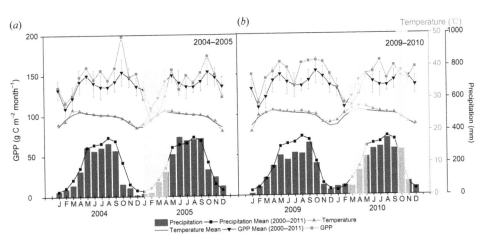

Figure 8.   Monthly GPP, temperature, and precipitation averaged over the LMB in (a) 2004–2005 and (b) 2009–2010. The orange box denotes the period experiencing GPP reduction. For the 2000 –2011 GPP mean, the error bars denote mean ± standard error.

back to May 2009, and the annual total precipitation in 2009 was about 15.6% lower than the 12 year mean. Although precipitation increased in January and February 2010, the monthly total precipitation between March and December was 14.6% lower than the 12 year mean. Moreover, the monthly mean temperatures from January to May 2010 were higher than the 12 year mean (1.3°C, 1.2°C, 1.1°C, 1.1°C, and 1.3°C higher in January, February, March, April, and May, respectively). The regional drought in 2009 and 2010 has been attributed to the weakness of the 2009 southwest monsoon and its early withdrawal (MRC 2010). Zhou et al. (2011) found that the upper reaches of the Mekong River Basin experienced a lack of precipitation from November 2009 to January 2010, which caused most rivers to fall to their lowest levels on record (MRC 2010); this relative lack of available water affected groundwater, soil moisture, and the growth of plants in the LMB.

The meteorological differences between the two years likely contributed to the observed differences in the reduction of GPP. The decrease of GPP in 2005 was more severe than that in 2010 in terms of both intensity and duration. The reduction of GPP in both 2005 and 2010 occurred primarily in winter and spring. In 2005, GPP decreased by 9.1%, 13.6%, 17.5%, and 12.8% in January, February, March, and April, respectively. Conversely, in 2010, GPP reduction occurred during March and April, with decreases of 8.9% and 12.7%, respectively. The shorter duration and intensity of these GPP changes in 2010 can be explained by the fact that the stress conditions in 2010 were alleviated considerably by the exceptionally high precipitation in January and February, when precipitation increased by 68.4% and 12.9%, respectively. Our result is in accordance with other studies that have demonstrated that winter and spring droughts can affect productivity by shortening the length of the growing season and suppressing canopy development and peak leaf area, thus leading to a decline in annual net carbon uptake (Noormets et al. 2009).

As Cambodia was affected most severely by drought, we selected Cambodia as an example to further analyse the seasonal characteristics of the droughts during 2005 and 2010 (Figure 9). In 2005, 92.3% (7.7%) of Cambodia was hit by severe drought (mild drought); conversely, in 2010, 69.2% of Cambodia suffered from mild drought, with no severe drought recorded. Because drought in both years was either predominantly severe (i.e. in 2005) or entirely mild (i.e. in 2010), we simply calculated mean GPP for drought conditions (PDSI < −1) and did not separate our results into two levels of drought in the following analysis. Our results demonstrate that the drought in 2005 caused larger reductions in GPP than that in 2010 for all vegetation types. In 2005 (2010), drought occurred primarily in winter and early spring (spring and summer). However, the drought in 2005 can be traced back to 2004 and even to 2003 in some regions, when drought lowered the normal water levels and river flows (Te 2007). For all vegetation types, the reduction in GPP in 2005 began from January to July and lasted for 7 months; thus, these reductions persisted for a longer period than those in 2010 (January–June; 6 months). During January–April 2005, the decline of evergreen woodland and shrublands was the greatest (215.4 g C m^{-2}), followed by evergreen forests (211.9 g C m^{-2}), deciduous forests/woodland and shrublands (154.1 g C m^{-2}), croplands/vegetation mosaics (126.4 g C m^{-2}), and single-season croplands (64.0 g C m^{-2}). During March and May in 2010, GPP declined by 68.9, 62.4, 44.7, and 33.9 g C m^{-2} in croplands/vegetation mosaics, deciduous forests/woodland and shrublands, evergreen woodland and shrublands, and single-season croplands, respectively. In contrast to other vegetation types, the variation in GPP in 2010 for single-season croplands was very close to the 12 year mean; we attribute the less-pronounced effects of drought in this instance to human intervention (such as irrigation). The 2010 drought was

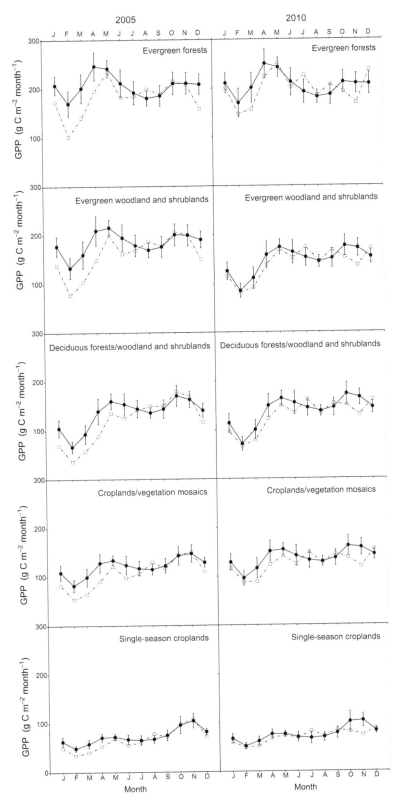

Figure 9.   Monthly variations in regional mean GPP under drought condition (PDSI < −1) for 2005, 2010, and 2000−2011 average in Cambodia. The dashed and solid lines denote the 12 year mean and monthly GPP, respectively. For the 2000−2011 mean, the error bars denote mean ± standard error.

characterized primarily by mild drought, with no areas affected by severe drought. However, severe drought affected 92.3% of the study area in 2005, inducing larger reductions in GPP for one-season croplands in 2005 relative to 2010. Our results imply that, although human practices may mitigate the effects of drought on croplands to some degree, drought mitigation measures cannot prevent reductions in plant yield when drought becomes extreme (Shimizu, Masumoto, and Pham 2006).

To detect the impact of drought on different types of crops, we further analysed the variation in GPP between different vegetation types in the LMB within Vietnam (including the majority of the Mekong Delta) (Figure 10). The Mekong Delta is a large rural area, dominated by aquaculture and rice paddy cultivation. A dense and complex hydrologic network of man-made canals, dikes, and sluices used for transport and irrigation (Leinenkugel, Esch, and Kuenzer 2011) allows the cultivation of two-season (19.2%) and even three-season (7.8%) croplands. Only a small area of Vietnam within the LMB experienced drought (PDSI < −1) in 2010; therefore, we focused our analysis for this region in 2005. Large differences in the reduction of NPP can be observed between forests, woodlands and shrublands, and croplands in response to the 2005 drought (Figure 10). The GPP variations obtained for two-season and three-season croplands are approximately within the 12 year standard deviation, whereas evergreen forests and evergreen woodland and shrublands experienced large GPP reductions during January–July. Compared to croplands/vegetation mosaics in Vietnam and single-season croplands in Cambodia in 2005, two-season and three-season croplands in Vietnam were impacted less by the drought in 2005. In January, the GPP values of two-season and three-season croplands were 152.1 and 159.2 g C m^{-2} month^{-1}, respectively; these values are 18.6 and 20.9 g C m^{-2} month^{-1} higher than the 12 year mean. It appears that farming practices can reduce the impact of drought, particularly through the use of different rice varieties in different harvest cycles. Most paddy fields in the Mekong Delta are planted with double irrigated-rice crops, double rain-fed rice crops, or triple irrigated-rice crops growing in different seasons (Sakamoto et al. 2006); therefore, the

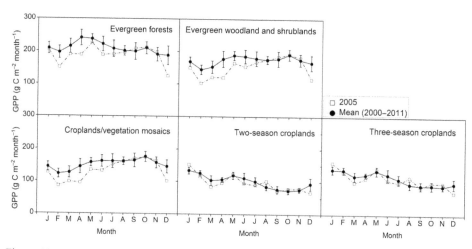

Figure 10. Monthly variations in regional mean GPP under drought conditions (PDSI < −1) for 2005, and the 2000–2011 average in the part of Vietnam within the LMB (including the majority of the Mekong Delta). The dashed and solid lines denote the 12 year mean and monthly GPP, respectively. For the 2000–2011 mean, the error bars denote mean ± standard error.

winter and spring drought in 2005 likely influenced only a short part of the total growing period for these croplands types.

4. Conclusions

In the present study, we investigated the impact of droughts on vegetation productivities in the LMB during 2000–2011. Several drought events occurred in the LMB during 2000–2011: mild droughts were observed in 2003, 2004, 2006, and 2007, whereas two severe drought events occurred in 2005 and 2010. Cambodia and Lao PDR were the countries affected most severely in 2005 and 2010, with PDSI values of –2.5 and –1.8, respectively. In terms of drought intensity, the drought in 2005 was more severe than that in 2010 and had a more significant impact on vegetation productivity in the LMB. For the entire study region, large reductions in NPP were found during both 2005 (14.7%) and 2010 (8.4%). The lack of precipitation during June 2004 and May 2005 resulted in GPP reduction $(66.5 \text{ g C m}^{-2})$ during January–April 2005. Similarly, in 2010, the reduction in GPP $(29.1 \text{ g C m}^{-2})$ between March and May could be traced back to a lack of precipitation that commenced in May 2009 and lasted until September 2010 (excluding January and February in 2010). The longer duration and larger deficit of precipitation in 2005 resulted in more severe drought in this year; thus, larger GPP reductions were observed in 2005 than in 2010.

Variations in NPP differed between vegetation types, and severe drought (PDSI < –2) exerted a greater impact on NPP than mild drought (–2 < PDSI < –1). In the LMB, the NPP of forests, woodlands, and shrublands exhibited larger reductions than that of croplands, which we attribute to the wider adoption of irrigation and other water-control measures in croplands, such as those in Cambodia and the Vietnam Mekong Delta. Furthermore, during the 2005 drought, two-season and three-season croplands in the Mekong Delta were impacted less by drought than single-season croplands in Cambodia and croplands/vegetation mosaics in Vietnam, likely owing to the more extensive irrigation and adoption of multiple harvest cycles in these two-and three-season croplands.

Although there are some limitations imposed by the quality of the MODIS productivity product, our study still provides an insight into how extreme climate may affect tropical and subtropical vegetation ecosystems in the LMB, a region that is critical to global bioecology. Our results suggest that, in regions in which conditions are appropriate for multi-season harvest (e.g. sufficient water supplies), local farmers could change rice varieties (i.e. cultivate two-season and three-season croplands) and adapt their harvesting and sowing cycles where applicable to reduce the effects of short periods of drought. Furthermore, improvements in irrigation technology may contribute to the mitigation of drought impacts in regions where irrigation is possible.

Acknowledgements
The authors would like to thank the anonymous reviewers for their valuable comments and suggestions for revising and improving the article.

Funding
This work was supported by International Cooperation and Exchanges NSFC [grant number 41120114001]; National Natural Science Foundation of China under [grant number 41271372].

References

ADB (Asian Development Bank). 2009. *The Economics of Climate Change in Southeast Asia: A Regional Review*. Jakarta: ADB. Accessed November 14, 2013. http://www.adb.org/publications/economics-climate-change-southeast-asia-regional-review

Allen, C. D., A. K. Macalady, H. Chenchouni, D. Bachelet, N. McDowell, M. Vennetier, T. Kitzberger, A. Rigling, D. D. Breshears, E. H. Hogg, P. Gonzalez, R. Fensham, Z. Zhang, J. Castro, N. Demidova, J. H. Lim, G. Allard, S. W. Running, and N. Cobb. 2010. "A Global Overview of Drought and Heat-Induced Tree Mortality Reveals Emerging Climate Change Risks for Forests." *Forest Ecology and Management* 259: 660–684. doi:10.1016/j.foreco.2009.09.001.

Alley, W. M. 1984. "The Palmer Drought Severity Index: Limitations and Assumptions." *Journal of Climate and Applied Meteorology* 23: 1100–1109. doi:10.1175/1520-0450(1984)0232.0.CO;2.

Asner, G. P., and A. Alencar. 2010. "Drought Impacts on the Amazon Forest: The Remote Sensing Perspective." *New Phytologist* 187: 569–578. doi:10.1111/j.1469-8137.2010.03310.x.

Bhalme, H. N., and D. A. Mooley. 1980. "Large-Scale Droughts/Floods and Monsoon Circulation." *Monthly Weather Review* 108: 1197–1211. doi:10.1175/1520-0493(1980)1082.0.CO;2.

Buckley, B. M., K. Palakit, K. Duangsathaporn, P. Sanguantham, and P. Prasomsin. 2007. "Decadal Scale Droughts Over Northwestern Thailand Over the Past 448 Years: Links to the Tropical Pacific and Indian Ocean Sectors." *Climate Dynamics* 29: 63–71. doi:10.1007/s00382-007-0225-1.

Chen, G., H. Tian, C. Zhang, M. Liu, W. Ren, W. Zhu, A. H. Chappelka, S. A. Prior, and G. B. Lockaby. 2012. "Drought in the Southern United States Over the 20th Century: Variability and its Impacts on Terrestrial Ecosystem Productivity and Carbon Storage." *Climatic Change* 114: 379–397. doi:10.1007/s10584-012-0410-z.

Ciais, P., M. Reichstein, N. Viovy, A. Granier, J. Ogée, V. Allard, M. Aubinet, N. Buchmann, C. Bernhofer, A. Carrara, F. Chevallier, N. De Noblet, A. D. Friend, P. Friedlingstein, T. Grünwald, B. Heinesch, P. Keronen, A. Knohl, G. Krinner, D. Loustau, G. Manca, G. Matteucci, F. Miglietta, J. M. Ourcival, D. Papale, K. Pilegaard, S. Rambal, G. Seufert, J. F. Soussana, M. J. Sanz, E. D. Schulze, T. Vesala, and R. Valentini. 2005. "Europe-Wide Reduction in Primary Productivity Caused by the Heat and Drought in 2003." *Nature* 437: 529–533. doi:10.1038/nature03972.

Dai, A., K. E. Trenberth, and T. Qian. 2004. "A Global Dataset of Palmer Drought Severity Index for 1870-2002: Relationship With Soil Moisture and Effects of Surface Warming." *Journal of Hydrometeorology* 5: 1117–1130. doi:10.1175/JHM-386.1.

Domonkos, P., S. Szalai, and J. Zoboki. 2001. "Analysis of Drought Severity Using PDSI and SPI Indices." *IdŰjárás* 105: 93–107.

Dos Santos, R., and A. R. Pereira. 1999. "Palmer Drought Severity Index for Western Sao Paulo State, Brazil." *Rev. Bras. Agrometeor* 7: 139–145.

Dudgeon, D. 2000. "Large-Scale Hydrological Changes in Tropical Asia: Prospects for Riverine Biodiversity: The Construction of Large Dams will have an Impact on the Biodiversity of Tropical Asian Rivers and Their Associated Wetlands." *Bioscience* 50: 793–806. doi:10.1641/0006-3568(2000)050[0793:LSHCIT]2.0.CO;2.

Eastham, J., F. Mpelasoka, M. Mainuddin, C. Ticehurst, P. Dyce, G. Hodgson, R. Ali, and M. Kirby. 2008. *Mekong River Basin Water Resources Assessment: Impacts of Climate Change*. Canberra: CSIRO: Water for a Healthy Country National Research Flagship.

FAO (Food and Agriculture Organization of the United Nations). 2007. *State of the World's Forests*: 2007. Rome: FAO.

Field, C. B., M. J. Behrenfeld, J. T. Randerson, and P. Falkowski. 1998. "Primary Production of the Biosphere: Integrating Terrestrial and Oceanic Components." *Science* 281: 237–240. doi:10.1126/science.281.5374.237.

Fu, C., and G. Wen. 1999. "Variation of Ecosystems Over East Asia in Association With Seasonal, Interannual and Decadal Monsoon Climate Variability." *Climatic Change* 43: 477–494. doi:10.1023/A:1005471600483.

Gilgen, A. K., and N. Buchmann. 2009. "Response of Temperate Grasslands at Different Altitudes to Simulated Summer Drought Differed But Scaled with Annual Precipitation." *Biogeosciences Discussions* 6: 5217–5250. doi:10.5194/bgd-6-5217-2009.

Giuffrida, A., and M. Conte. 1989. "Long Term Evolution of the Italian Climate Outlined by Using the Standardized Anomaly Index (SAI)." *Conference on Climate and Water* 1: 197–208.

Hanson, P. J., and J. F. Weltzin. 2000. "Drought Disturbance from Climate Change: Response of United States Forests." *Science of the Total Environment* 262: 205–220. doi:10.1016/S0048-9697(00)00523-4.

Heim, R. R. 2000. "Drought Indices: A Review." In *Drought: A Global Assessment*, edited by Doald A. Wihite, 159–167. London: Routledge.

Hirata, R., N. Saigusa, S. Yamamoto, Y. Ohtani, R. Ide, J. Asanuma, M. Gamo, T. Hirano, H. Kondo, Y. Kosugi, S. Li, Y. Nakai, K. Takagi, M. Tani, and H. Wang. 2008. "Spatial Distribution of Carbon Balance in Forest Ecosystems Across East Asia." *Agricultural and Forest Meteorology* 148: 761–775. doi:10.1016/j.agrformet.2007.11.016.

IPCC. 2007. *Climate Change 2007: Impacts, Adaptation and Vulnerability: Working Group II Contribution to the Fourth Assessment Report of the IPCC Intergovernmental Panel on Climate Change*, 976 p., Vol. 4. Cambridge: Cambridge University Press.

Karl, T. R., and D. R. Easterling. 1999. "Climate Extremes: Selected Review and Future Research Directions." In *Weather and Climate Extremes*, edited by T. R. Karl, N. Nicholls, and A. Ghazi, 309–325. Netherlands: Springer.

Kobayashi, H., T. Matsunaga, and A. Hoyano. 2005. "Net Primary Production in Southeast Asia Following a Large Reduction in Photosynthetically Active Radiation Owing To Smoke." *Geophysical Research Letters* 32. doi:10.1029/2004GL021704.

Kuenzer, C., I. Campbell, M. Roch, P. Leinenkugel, V. Q. Tuan, and S. Dech. 2012. "Understanding the Impact of Hydropower Developments in the Context of Upstream – Downstream Relations in the Mekong River Basin." *Sustainability Science* 8: 565–584. doi:10.1007/s11625-012-0195-z.

Kuenzer, C., H. Guo, J. Huth, P. Leinenkugel, X. Li, and S. Dech. 2013. "Flood Mapping and Flood Dynamics of the Mekong Delta: ENVISAT-ASAR-WSM Based Time Series Analyses." *Remote Sensing* 5: 687–715. doi:10.3390/rs5020687.

Kwon, H., E. Pendall, B. E. Ewers, M. Cleary, and K. Naithani. 2008. "Spring Drought Regulates Summer Net Ecosystem CO_2 Exchange in a Sagebrush-Steppe Ecosystem." *Agricultural and Forest Meteorology* 148: 381–391. doi:10.1016/j.agrformet.2007.09.010.

Kwon, Y., and C. P. Larsen. 2013. "Effects of Forest Type and Environmental Factors on Forest Carbon Use Efficiency Assessed Using MODIS and FIA Data Across the Eastern USA." *International Journal of Remote Sensing* 34: 8425–8448. doi:10.1080/01431161.2013.838711.

Leinenkugel, P., T. Esch, and C. Kuenzer. 2011. "Settlement Detection and Impervious Surface Estimation in the Mekong Delta Using Optical and SAR Remote Sensing Data." *Remote Sensing of Environment* 115: 3007–3019. doi:10.1016/j.rse.2011.06.004.

Leinenkugel, P., C. Kuenzer, and S. Dech. 2013. "Comparison and Enhancement of MODIS Cloud Mask Products for Southeast Asia." *International Journal of Remote Sensing* 34: 2730–2748. doi:10.1080/01431161.2012.750037.

Leinenkugel, P., C. Kuenzer, N. Oppelt, and S. Dech. 2013. "Characterisation of Land Surface Phenology and Land Cover Based on Moderate Resolution Satellite Data in Cloud Prone Areas – A Novel Product for the Mekong Basin." *Remote Sensing of Environment* 136: 180–198. doi:10.1016/j.rse.2013.05.004.

Lewis, S. L., P. M. Brando, O. L. Phillips, G. M. van der Heijden, and D. Nepstad. 2011. "The 2010 Amazon Drought." *Science* 331: 554. doi:10.1126/science.1200807.

Masumoto, T., P. T. Hai, and K. Shimizu. 2008. "Impact of Paddy Irrigation Levels on Floods and Water Use in the Mekong River Basin." *Hydrological Processes* 22: 1321–1328. doi:10.1002/hyp.6941.

Matsushita, B., and M. Tamura. 2002. "Integrating Remotely Sensed Data with an Ecosystem Model to Estimate Net Primary Productivity in East Asia." *Remote Sensing of Environment* 81: 58–66. doi:10.1016/S0034-4257(01)00331-5.

Matsushita, B., M. Xu, J. Chen, S. Kameyama, and M. Tamura. 2004. "Estimation of Regional Net Primary Productivity (NPP) Using a Process-Based Ecosystem Model: How Important is the Accuracy of Climate Data?" *Ecological Modelling* 178: 371–388. doi:10.1016/j.ecolmodel.2004.03.012.

McKee, T. B., N. J. Doesken, and J. Kleist. 1995. "Drought Monitoring with Multiple Time Scales." Ninth Conference on Applied Climatology, American Meteorological Society, Boston, MA, January 15–20.

Mika, J., S. Z. Horváth, L. Makra, and Z. Dunkel. 2005. "The Palmer Drought Severity Index (PDSI) as an Indicator of Soil Moisture." *Physics and Chemistry of the Earth, Parts A/B/C* 30: 223–230. doi:10.1016/j.pce.2004.08.036.

107

Mohammat, A., X. Wang, X. Xu, L. Peng, Y. Yang, X. Zhang, R. B. Myneni, and S. Piao. 2012. "Drought and Spring Cooling Induced Recent Decrease in Vegetation Growth in Inner Asia." *Agricultural and Forest Meteorology* 178: 21–30. doi:10.1016/j.agrformet.2012.09.014.

MRC (Mekong River Commission). 2003. *State of the Basin Report 2003: Executive Summary*, 50 p. Phnom Penh: MRC.

MRC (Mekong River Commission). 2005. *Lower Mekong Basin Drought Study: Analysis, Forecasting, Planning and Management*. Vientiane: MRC.

MRC (Mekong River Commission). 2009. *Adaptation to Climate Change in the Countries of the Lower Mekong Basin: Regional Synthesis Report*, 89 p. Vietnam: MRC.

MRC (Mekong River Commission). 2010. *State of the Basin Report 2010*. Vientiane: MRC.

MRC (Mekong River Commission). 2011. *Flood Situation Report 2011*, 57 p. Phnom Penh: MRC., 57 pp.

Mu, Q. Z., M. S. Zhao, J. S. Kimball, N. G. McDowell, and S. W. Running. 2013. "A Remotely Sensed Global Terrestrial Drought Severity Index." *Bulletin of the American Meteorological Society* 94 (1): 83–98. doi:10.1175/BAMS-D-11-00213.1.

Naeimi, V., P. Leinenkugel, D. Sabel, W. Wagner, H. Apel, and C. Kuenzer. 2013. "Evaluation of Soil Moisture Retrieval from the ERS and Metop Scatterometers in the Lower Mekong Basin." *Remote Sensing* 5: 1603–1623. doi:10.3390/rs5041603.

Nesbitt, H., R. Johnston, and M. Solieng. 2003. "Mekong River Water: Will River Flows Meet Future Agriculture Needs in the Lower Mekong Basin?" *Water in Agriculture* 116: 86–104.

Noormets, A., S. G. McNulty, J. L. DeForest, G. Sun, Q. Li, and J. Chen. 2008. "Drought During Canopy Development has Lasting Effect on Annual Carbon Balance in a Deciduous Temperate Forest." *New Phytologist* 179: 818–828. doi:10.1111/j.1469-8137.2008.02501.x.

Ntale, H. K., and T. Y. Gan. 2003. "Drought Indices and Their Application to East Africa." *International Journal of Climatology* 23: 1335–1357. doi:10.1002/joc.931.

Palmer, W. C. 1965. *Meteorological Drought*. Washington, DC: US Department of Commerce, Weather Bureau.

Pantuwan, G., S. Fukai, M. Cooper, S. Rajatasereekul, and J. C. O. Toole. 2002. "Yield Response of Rice (Oryza Sativa L.) Genotypes to Different Types of Drought Under Rainfed Lowlands: Part 1. Grain Yield and Yield Components." *Field Crops Research* 73: 153–168. doi:10.1016/S0378-4290(01)00187-3.

Patra, P. K., M. Ishizawa, S. Maksyutov, T. Nakazawa, and G. Inoue. 2005. "Role of Biomass Burning and Climate Anomalies for Land-Atmosphere Carbon Fluxes Based on Inverse Modeling of Atmospheric CO2." *Global Biogeochemical Cycles* 19 (3). doi:10.1029/2004GB002258.

Paw, J. N., and C. Thia-Eng. 1991. "Climate Changes and Sea Level Rise: Implications on Coastal Area Utilization and Management in South-East Asia." *Ocean and Shoreline Management* 15: 205–232. doi:10.1016/0951-8312(91)90043-2.

Pei, F., X. Li, X. Liu, and C. Lao. 2013, "Assessing the Impacts of Droughts on Net Primary Productivity in China." *Journal of Environmental Management*. doi:10.1016/j.jenvman.2012.10.031.

Pennington, D. D., and S. L. Collins. 2007. "Response of an Aridland Ecosystem to Interannual Climate Variability and Prolonged Drought." *Landscape Ecology* 22: 897–910. doi:10.1007/s10980-006-9071-5.

Pereira, J. S., J. A. Mateus, L. M. Aires, G. Pita, C. Pio, J. S. David, V. Andrade, J. Banza, T. S. David, T. A. Paço, and A. Rodrigues. 2007. "Net Ecosystem Carbon Exchange in Three Contrasting Mediterranean Ecosystems – The Effect of Drought." *Biogeosciences* 4: 791–802. doi:10.5194/bg-4-791-2007.

Peters, A. J., E. A. Walter-Shea, L. Ji, A. Vina, M. Hayes, and M. D. Svoboda. 2002. "Drought Monitoring with NDVI-Based Standardized Vegetation Index." *Photogrammetric Engineering and Remote Sensing* 68: 71–75.

Potter, C., S. Klooster, C. Hiatt, V. Genovese, and J. C. Castilla-Rubio. 2011. "Changes in the Carbon Cycle of Amazon Ecosystems During the 2010 Drought." *Environmental Research Letters* 6: 034024. doi:10.1088/1748-9326/6/3/034024.

Propastin, P., A. Ibrom, A. Knohl, and S. Erasmi. 2012. "Effects of Canopy Photosynthesis Saturation on the Estimation of Gross Primary Productivity from MODIS Data in a Tropical Forest." *Remote Sensing of Environment* 121: 252–260. doi:10.1016/j.rse.2012.02.005.

Rambal, S., J. M. Ourcival, R. Joffre, F. Mouillot, Y. Nouvellon, M. Reichstein, and A. Rocheteau. 2003. "Drought Controls Over Conductance and Assimilation of a Mediterranean Evergreen Ecosystem: Scaling from Leaf to Canopy." *Global Change Biology* 9: 1813–1824. doi:10.1111/j.1365-2486.2003.00687.x.

Räsänen, T. A., and M. Kummu. 2012. "Spatiotemporal Influences of ENSO on Precipitation and Flood Pulse in the Mekong River Basin." *Journal of Hydrology* 476: 154–168. doi:10.1016/j.jhydrol.2012.10.028.

Reeves, M. C., M. Zhao, and S. W. Running. 2005. "Usefulness and Limits of MODIS GPP for Estimating Wheat Yield." *International Journal of Remote Sensing* 26: 1403–1421. doi:10.1080/01431160512331326567.

Rienecker, M. M., M. J. Suarez, R. Gelaro, R. Todling, J. Bacmeister, E. Liu, M. G. Bosilovich, S. D. Schubert, L. Takacs, G. Kim, S. Bloom, J. Chen, D. Collins, A. Conaty, A. da Silva, W. Gu, J. Joiner, R. D. Koster, R. Lucchesi, A. Molod, T. Owens, S. Pawson, P. Pegion, C. R. Redder, R. Reichle, F. R. Robertson, A. G. Ruddick, M. Sienkiewicz, and J. Woollen. 2011. "MERRA: NASA's Modern-Era Retrospective Analysis for Research and Applications." *Journal of Climate* 24: 3624–3648. doi:10.1175/JCLI-D-11-00015.1.

Rosenzweig, C., A. Iglesias, X. B. Yang, P. R. Epstein, and E. Chivian. 2001. "Climate Change and Extreme Weather Events; Implications for Food Production, Plant Diseases, and Pests." *Global Change & Human Health* 2: 90–104. doi:10.1023/A:1015086831467.

Saigusa, N., K. Ichii, H. Murakami, R. Hirata, J. Asanuma, H. Den, S. Han, R. Ide, S. Li, T. Ohta, T. Sasai, S. Wang, and G. Yu. 2010. "Impact of Meteorological Anomalies in the 2003 Summer on Gross Primary Productivity in East Asia." *Biogeosciences* 7: 641–655. doi:10.5194/bg-7-641-2010.

Sakamoto, T., N. Van Nguyen, H. Ohno, N. Ishitsuka, M. Yokozawa. 2006. "Spatio–Temporal Distribution of Rice Phenology and Cropping Systems in the Mekong Delta with Special Reference to the Seasonal Water Flow of the Mekong and Bassac Rivers." *Remote Sensing of Environment* 100: 1–16. doi:10.1016/j.rse.2005.09.007.

Sakamoto, T., C. Van Phung, A. Kotera, K. D. Nguyen, and M. Yokozawa. 2009. "Analysis of Rapid Expansion of Inland Aquaculture and Triple Rice-Cropping Areas in a Coastal Area of the Vietnamese Mekong Delta Using MODIS Time-Series Imagery." *Landscape and Urban Planning* 92: 34–46. doi:10.1016/j.landurbplan.2009.02.002.

Samanta, A., S. Ganguly, H. Hashimoto, S. Devadiga, E. Vermote, Y. Knyazikhin, R. R. Nemani, and R. B. Myneni. 2010. "Amazon Forests did not Green-Up During the 2005 Drought." *Geophysical Research Letters* 37. doi:10.1029/2009GL042154.

Scott, R. L., G. D. Jenerette, D. L. Potts, and T. E. Huxman. 2009. "Effects of Seasonal Drought on Net Carbon Dioxide Exchange from a Woody-Plant-Encroached Semiarid Grassland." *Journal of Geophysical Research: Biogeosciences (2005–2012)* 114 (G4). doi:10.1029/2008JG000900.

Shiba, S. S. T., and A. Apan. 2011. "Analysing the Effect of Drought on Net Primary Productivity of Tropical Rainforests in Queensland Using MODIS Satellite Imagery." *Proceeding of the Surveying& Spatial Sciences Biennial Conference* 2011: 97–110.

Shimizu, K., T. Masumoto, and T. H. Pham. 2006. "Factors Impacting Yields in Rain-Fed Paddies of the Lower Mekong River Basin." *Paddy and Water Environment* 4: 145–151. doi:10.1007/s10333-006-0041-y.

Son, N. T., C. F. Chen, C. R. Chen, L. Y. Chang, and V. Q. Minh. 2012. "Monitoring Agricultural Drought in the Lower Mekong Basin Using MODIS NDVI and Land Surface Temperature Data." *International Journal of Applied Earth Observation and Geoinformation* 18: 417–427. doi:10.1016/j.jag.2012.03.014.

Southavilay, T., and T. Castrén. 1999. "Timber Trade and Wood Flow Study-Lao PDR." Regional Environmental Technical Assistance 5771. Poverty Reduction and Environmental Management in Remote Greater Mekong Subregion (GMS) Watersheds Project (Phase I). Accessed November 14, 2013. http://www.mekonginfo.org/assets/midocs/0002916-environment-timber-trade-and-wood-flow-study-lao-pdr.pdf

Sovacool, B. K., A. L. D. Agostino, A. Rawlani, and H. Meenawat. 2012. "Improving Climate Change Adaptation in Least Developed Asia." *Environmental Science & Policy* 21: 112–125. doi:10.1016/j.envsci.2012.04.009.

Taylor, D., P. Saksena, P. G. Sanderson, and K. Kucera. 1999. "Environmental Change and Rain Forests on the Aunda Shelf of Southeast Asia: Drought, Fire and the Biological Cooling of

Biodiversity Hotspots." *Biodiversity & Conservation* 8: 1159–1177. doi:10.1023/A:1008952428475.

Te, N. 2007. "Drought Management in the Lower Mekong Basin." Paper presentrd at 3rd South Asia Water Forum, Kuala Lumpur, October 22–26.

Tian, H., J. M. Melillo, D. W. Kicklighter, S. Pan, J. Liu, A. D. McGuire, and B. Moore III. 2003. "Regional Carbon Dynamics in Monsoon Asia and Its Implications for the Global Carbon Cycle." *Global and Planetary Change* 37: 201–217.

Turner, D. P., W. D. Ritts, M. Zhao, S. A. Kurc, A. L. Dunn, S. C. Wofsy, E. E. Small, and S. W. Running. 2006. "Assessing Inter-Annual Variation in MODIS-Based Estimates of Gross Primary Production." *IEEE Transactions on Geoscience and Remote Sensing* 44: 1899–1907. doi:10.1109/TGRS.2006.876027.

Van Rooy, M. P. 1965. "A Rainfall Anomaly Index Independent of Time and Space." *Notos* 14: 43–48.

Westerling, A. L., H. G. Hidalgo, D. R. Cayan, and T. W. Swetnam. 2006. "Warming and Earlier Spring Increase Western US Forest Wildfire Activity." *Science* 313: 940–943. doi:10.1126/science.1128834.

Wilhite, D. A. 2000. "Drought as a Natural Hazard: Concepts and Definitions." *Drought, a Global Assessment* 1: 3–18.

Xue, Z., J. P. Liu, and Q. Ge. 2011. "Changes in Hydrology and Sediment Delivery of the Mekong River in the Last 50 Years: Connection to Damming, Monsoon, and ENSO." *Earth Surface Processes and Landforms* 36: 296–308. doi:10.1002/esp.2036.

Zeng, H. Q., Q. J. Liu, Z. W. Feng, X. K. Wang, and Z. Q. Ma. 2008. "Modeling the Interannual Variation and Response to Climate Change Scenarios in Gross and Net Primary Productivity of Pinus Elliottii Forest in Subtropical China." *Acta Ecologica Sinica* 28: 5314–5321. doi:10.1016/S1872-2032(09)60008-1.

Zhang, F. M., W. M. Ju, J. M. Chen, S. Q. Wang, G. R. Yu, and S. J. Han. 2012. "Characteristics of Terrestrial Ecosystem Primary Productivity in East Asia Based on Remote Sensing and Process-Based Model." *Ying yong sheng tai xue bao= The Journal of Applied Ecology/Zhongguo sheng tai xue xue hui, Zhongguo ke xue yuan shenyang ying yong sheng tai yan jiu suo zhu ban* 23: 307.

Zhang, L., J. Xiao, J. Li, K. Wang, L. Lei, and H. Guo. 2012. "The 2010 Spring Drought Reduced Primary Productivity in Southwestern China." *Environmental Research Letters* 7: 045706. doi:10.1088/1748-9326/7/4/045706.

Zhang, X., M. Goldberg, D. Tarpley, M. A. Friedl, J. Morisette, F. Kogan, and Y. Yu. 2010. "Drought-Induced Vegetation Stress in Southwestern North America." *Environmental Research Letters* 5: 024008. doi:10.1088/1748-9326/5/2/024008.

Zhao, M., F. A. Heinsch, R. R. Nemani, and S. W. Running. 2005. "Improvements of the MODIS Terrestrial Gross and Net Primary Production Global Data Set." *Remote Sensing of Environment* 95: 164–176. doi:10.1016/j.rse.2004.12.011.

Zhao, M., and S. W. Running. 2010. "Drought-Induced Reduction in Global Terrestrial Net Primary Production from 2000 Through 2009." *Science* 329: 940–943. doi:10.1126/science.1192666.

Zhao, M., S. W. Running, and R. R. Nemani. 2006. "Sensitivity of Moderate Resolution Imaging Spectroradiometer (MODIS) Terrestrial Primary Production to the Accuracy of Meteorological Reanalyses." *Journal of Geophysical Research: Biogeosciences (2005–2012)* 111. doi:10.1029/2004JG000004.

Zhou, J., Z. Zhang, G. Sun, X. Fang, T. Zha, S. McNulty, J. Chen, Y. Jin, and A. Noormets. 2013. "Response of Ecosystem Carbon Fluxes to Drought Events in a Poplar Plantation in Northern China." *Forest Ecology and Management.* doi:10.1016/j.foreco.2013.01.007.

Zhou, T., C. Zhou, F. Yu, and Y. Zhao. 2011. "Spatial and Temporal Distribution Characteristics Analysis of Meteorological Drought in Lancang–Mekong River Basin." *Water Resources and Power* 6: 3.

Suitability of SAR imagery for automatic flood mapping in the Lower Mekong Basin

Felix Greifeneder, Wolfgang Wagner, Daniel Sabel, and Vahid Naeimi

Department for Geodesy and Geoinformation, Research Group for Remote Sensing, Vienna University of Technology, Vienna, Austria

Flood detection and inundation mapping are amongst the most important applications for remote-sensing data. Space-borne radar systems, synthetic aperture radar (SAR) in particular, and its application for waterbody mapping have recently been subject to research in many publications. Although very good results have been achieved with such data, in some cases automatic waterbody classification based on SAR data is not feasible. Factors influencing the applicability are, e.g., local environmental conditions, roughening of water surfaces due to wind, or the satellite observation geometry. In this study, a measure for the usability of SAR imagery for flood mapping was investigated. Additionally, a method for permanent waterbody mapping was introduced. The study is based on Envisat ASAR wide swath mode (150 m spatial resolution) data of the Mekong River Basin. For the usability measure, the concept of 'high-contrast tiles' was established, which allows an *a priori* estimation of the expected accuracy of a waterbody classifier. The SAR-based permanent waterbody map was used for the validation of the approach. It was found that, for the test site, the new SAR usability measure allows the identification of unsuitable scenes with a certainty of more than 90%. The method is expected to be very useful for near-real-time flood mapping applications where human interaction is neither desired nor feasible when large regions and large data volumes are considered.

Introduction

Flood detection and inundation mapping are two important remote-sensing applications. According to statistics from the International Strategy for Disaster Reduction (ISDR), flooding is the most frequent disaster in all parts of the world except for Africa and Oceania. Between the years 2001 and 2010, based on a yearly average, 45.6% of all natural disasters were floods (United Nations Strategy for Disaster Reduction 2011). It is the number one cause for losses from natural hazards – in 2011 approximately 106.5 million people were affected by flooding; the flood in Pakistan in 2010 alone caused an estimated damage of USD 9.6 billion (Juren and Khan 2010).

Synthetic aperture radar (SAR) data have been used successfully for waterbody classification, e.g. by Brakenridge et al. (2003), Bazi, Bruzzone, and Melgani (2005), Martinez and Letoan (2007), Martinis, Twele, and Voigt (2009), Hostache, Matgen, and Wagner (2012), and Kuenzer et al. (2013). These studies demonstrated that SAR is in general well suited to this task and provides some advantages over optical sensors: SAR sensors provide measurements independent of solar illumination and at low microwave frequencies (1–10 GHz); they are also nearly independent of atmospheric conditions such as cloud coverage (Woodhouse 2005). Consequently, one of the main advantages of SAR over optical data is its all-weather, day and night imaging capability. Owing to the side-looking viewing geometry of SAR systems and the high dielectric constant of water at microwave frequencies, a high percentage of radiation is specularly reflected away from the sensor over smooth water surfaces. Therefore water surfaces have in general a lower backscatter coefficient than most other ground-cover types (Dobson, Pierce, and Ulaby 1996). However, depending on local environmental conditions such as partially sub-merged vegetation or roughening of the water surface due to turbulent water, rain, and local wind conditions this assumption may be rendered invalid.

Many of the publications previously mentioned focus on manually selected SAR images with high contrast between water and non-water pixels. In case of a real flood event, such well-suited data might not be available. Especially in the case of systematic production and near-real-time data delivery where human interaction is not desired and may not be feasible, a method that automatically assesses the suitability of any given incoming SAR-scene would be highly valuable to reduce classification errors.

The main objective of this work was introducing a measure for the quantification of the usability of SAR data for automatic waterbody classification. The performance of the measure was assessed with the help of a reference image, which was derived from SAR time-series data.

Study region

The Mekong is the tenth largest river in the world. The basin of the Mekong River drains a total land area of 795,000 km^2 from the eastern watershed of the Tibetan Plateau to the Mekong Delta. The Mekong River flows approximately 4909 km through three provinces of China continuing through Myanmar, Laos, Thailand, Cambodia, and Vietnam before reaching the South China Sea.

This work focused on the Lower Mekong Basin (LMB), which is located within the boundaries of the countries Laos, Thailand, Vietnam, and Cambodia (Figure 1).

The climate of the Mekong River Basin ranges from temperate to tropical. Parts of the Upper Mekong Basin are snow covered in the winter, with some of the taller peaks of the Tibetan Plateau being glaciated. Melting snow from the plateau feeds the Mekong River's flow during the dry season, especially in the middle reaches. In the relatively lower elevations of the Yunnan province of China, the climate of the Mekong River Basin changes and the temperature gradually increases (Hoanh et al. 2003).

In the LMB, the wet season lasts from June to October; with the exception of two brief transition periods. The rest of the year is relatively dry, and is therefore considered as the dry season. Figure 2 shows a plot illustrating the average monthly precipitation and temperatures. The wet season results from the flow of moisture-laden air from the Indian Ocean in the summer. During the rest of the year, high-pressure systems over the Asian continent give rise to the dry season in the LMB (Hoanh et al. 2003).

As mentioned above, the LMB is subject to very strong seasonal variability. Naeimi et al. (2013) used soil moisture data derived from a space-borne scatterometer to illustrate the seasonal behaviour of soil moisture. Based on the general seasonality and the findings from Naeimi et al. (2013), the months December, January, February, and March were chosen to represent the dry season in the current study.

Figure 1. The Mekong River Basin. Lower Mekong Basin marked by the red boundary (Background map, © National Geographic, ESRI, DeLorme, NAVTEQ, UNEP-WCMC, USGS, NASA, ESA, METI, GEBCO, NOAA, IPC).

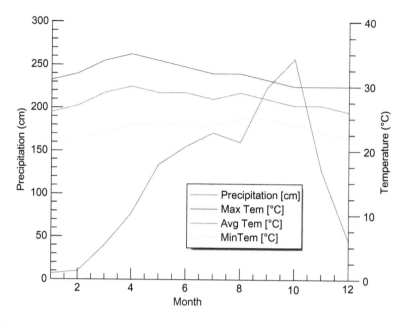

Figure 2. Average monthly precipitation and temperature for Phnom Penh in Cambodia.

Data

The data were acquired by the advanced synthetic aperture radar (ASAR) instrument on ESA's Envisat satellite. Envisat was launched on 1 March 2002, carrying a range of instruments with the main purpose of acquiring environmental data. The satellite was in operation until April 2012. The specifications of the used data are provided in Table 1.

Figure 3 shows the average number of ASAR images per location for different years and months. The number of measurements varies significantly between different time periods and regions, owing to the ASAR data acquisition strategy.

The validation of the automatically generated reference image was based on the Shuttle Radar Topography (SRTM) WaterBody Data set (Table 1). The SRTM WaterBody Data set (SWBD) was partly derived from radar data acquired during the February 2000 space shuttle mission and partly from Landsat 5 data. It depicts the state of waterbodies as of February 2000. The data set is distributed in a vector format by the United Stated Geological Survey (USGS). Waterbodies are included or excluded depending on their size. Lakes with a length greater than 600 m and wider than 183 m and rivers wider than 183 m for at least 600 m are included (Slater et al. 2006). Depiction of such a river ends if its width falls below 90 m. For the accuracy assessment the vector data set was rasterized to a sampling distance of approximately 75 m to match the spatial resolution of the ASAR WS data.

Table 1. Specifications of the data sets used in this study.

	Envisat ASAR wide swath mode (WS)	SRTM waterbody data set
Spatial resolution	150 m	~90 m
Polarization	HH/VV	–
Date of acquisition	2007–2011	2000
No. of images	696	1

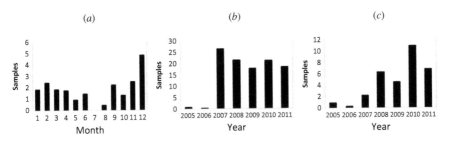

Figure 3.   The area of interest is covered by several SAR swaths. Therefore, the number of samples used for averaging varies from pixel to pixel. The figure shows the average number of samples for different time frames. (*a*) The average number of images per month in 2010; (*b*) Average number of images per year (all months). (*c*) Average number of images per year (only dry season).

Methodology

The approach for the automatic usability assessment of SAR imagery was based on parameters extracted from an Envisat ASAR WS data archive. Data acquired between 2007 and 2011 (696 images in total) were used to derive a temporal mean backscatter image, which acted as a benchmark to assess the quality of single-SAR images. Using a temporal mean backscatter, i.e. averaging over a long time period, reduces noise effects induced by speckle or wind on larger waterbodies. The result is a backscatter image with improved contrast between water and terrain, representing the mean state of permanent waterbodies.

The concept of high-contrast tiles was introduced as a method for the assessment of the usability of SAR images for waterbody classification. The algorithm identifies areas within the mean backscatter image, which are likely to contain water and no-water areas – so called high-contrast tiles. The selection of these areas was based on an improved version of the split-based approach, as had been introduced by Martinis, Twele, and Voigt (2009). The objective of this approach is to identify image subsets containing water and no-water pixels in an optimal distribution as well as sufficient contrast, which is necessary for a successful automatic waterbody classification. Using the knowledge of the locations of such areas, derived from the mean backscatter, the usability of any given SAR image can be assessed.

Figure 4 presents an overview of the processing chain. The data were geocoded, radiometrically calibrated, and subsequently re-sampled with bilinear interpolation to a common grid. To compensate for the effect of the local incidence angle on the radar backscatter, the data were normalized to a local incidence angle of 30° using the approach of Pathe et al. (2009). The resulting time-series database was used to generate a temporal mean backscatter image. This image was then used to produce the permanent waterbody map and to support the estimation of the suitability of single-SAR images using the concept of high-contrast tiles.

Definition of 'high-contrast tiles'

The temporal mean backscatter image was divided in overlapping sub-images with a dimension of 100 × 100 pixels and an overlap of 50%. Based on two selection criteria (Equations (1) and (2)) image tiles were qualified as high contrast and potentially containing water and non-water land-cover classes. Martinis, Twele, and Voigt (2009) introduced

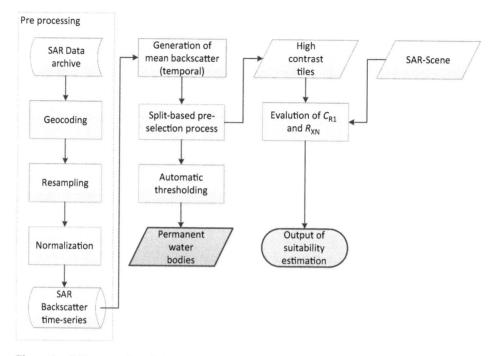

Figure 4.   SAR processing chain including preprocessing, permanent waterbody classification, and SAR-image evaluation.

a measure for the contrast within an image, on which the selection of appropriate sub-images can be based (Equation (1)).

$$C_{\mathrm{VX}} = \left| \frac{\mu_{\mathrm{w}}}{\sigma_{\mathrm{w}}} \right|, R_{\mathrm{XN}} = \left| \frac{\mu_{\mathrm{w}}}{\mu_{\mathrm{T}}} \right|. \tag{1}$$

Equation (1) shows the definition of two contrast parameters; μ_{w} and σ_{w} represent the mean and the standard deviation of the respective image tiles, and μ_{T} stands for the mean of the entire scene. According to Martinis, Twele, and Voigt, only tiles with C_{VX} values greater than or equal to 0.7 and with R_{XN} values between 0.4 and 0.9 were selected for further processing. However, it was found that the selection of tiles mainly depended on R_{XN}. C_{VX} hardly influenced the choice of tiles. Therefore, a new parameter was introduced to replace C_{VX}. C_{R1} (Equation (2)) is a measure for the image contrast calculated as the ratio between the tile's standard deviation (σ_{w}) and the range of all image grey-values (r_{T}). The threshold was set to 0.05, i.e. tiles with $C_{\mathrm{R1}} < 0.05$ do not qualify as high-contrast tiles. The location of each selected tile is stored.

$$C_{\mathrm{R1}} = \frac{\sigma_{\mathrm{w}}}{\gamma_{\mathrm{T}}}. \tag{2}$$

Based on the relative number of valid 'high-contrast tiles', in any given SAR image, the likely classification accuracy can be determined.

Usability assessment

The usability assessment approach is based on the assumption that the mean backscatter can be used as a benchmark to assess the usability of any given SAR scene for automatic waterbody classification. The evaluation of single-SAR images was based on high-contrast tiles, which were derived from the temporal mean backscatter image (Section 3.1) and stored in a database.

The first step is to restore the locations of high-contrast tiles from the database. For each tile, C_{VX} and C_{R1} (Equations (1) and (2)) were computed based on the backscatter values of the SAR image in question. The percentage of tiles satisfying the selection criteria acts as an indicator for the usability. The quantifier I_q (Equation (3)) was defined as the number of tiles (N_I) divided by the total number of high-contrast Tiles (N_T), as derived from the mean backscatter image, satisfying the criteria in the analysed SAR image. As an additional output a map, showing the distribution of selected tiles, can be generated. This allows further conclusions about the distribution of contrast (i.e. some areas might qualify for automatic thresholding whereas others within the same image do not, e.g. due to wind).

$$I_q = N_I/N_T. \tag{3}$$

Automatic waterbody mapping

For waterbody mapping, Otsu's (1979) automatic grey-level thresholding approach was adopted. This method is based on the assumption that, mostly due to surface roughness, backscatter from water surfaces is significantly lower than from land surfaces. Furthermore, it is assumed that the grey-level histogram resembles a bimodal distribution. The optimal threshold, separating the two histogram lobes, minimizes the combined spread (intra-class variance). Otsu's approach has been widely used for SAR image classification.

For any possible grey value, the 'goodness' is evaluated to determine the optimal threshold for image segmentation. L grey levels (Equation (5)) represent the pixels of a given image. Equations (4)–(6) describe the threshold selection. Using each possible grey value within the image as a threshold k, the pixels were assigned to two classes (foreground and background). Based on these classes the between-class variance was calculated for each threshold. The optimal location for k was found by maximizing the between-class variance. For this purpose the criterion function η was introduced (Equation (4)) as follows:

$$\eta(k) = \sigma_B^2(k)/\sigma_T^2. \tag{4}$$

In Equation (5), the definition of $\sigma_B{}^2$ and $\sigma_T{}^2$ is given as the between-class variance and the total variance of grey values, respectively:

$$\sigma_B^2 = \omega_0\omega_1(\mu_1 - \mu_0)^2 \quad \sigma_T^2 = \sum_{i=1}^{L}(i - \mu_T)^2 P_i, \tag{5}$$

where ω_0 and ω_1 (Equation (6)) are a measure for the probabilities of class occurrence for no-water and water classes, respectively, separated by the threshold value k. μ_0 and μ_1

represent the class mean levels. The total mean grey level of the original picture is expressed as μ_T.

$$\omega(k) = \sum_{i=1}^{k} P_i \mu(k) = \sum_{i=1}^{k} i P_i, \tag{6}$$

where P_i refers to the probability.

Equivalent number of looks

To assess the influence of the number of images used for the generation of the mean backscatter, the equivalent number of looks (ENL) was used as a measure for the level of noise within an image. ENL (Equation (7)) is defined as the ratio of the square of the mean of an image (μ^2) to the variance of an image (σ^2) (Gagnon and Jouan 1997).

$$\mathrm{ENL} = \frac{\mu^2}{\sigma^2}. \tag{7}$$

ENL is originally one parameter of multi-look SAR images, used to estimate the degree of averaging applied to the SAR measurements during data formation and post-processing. Multi-looking is performed to mitigate the noise-like effect of interference known as speckle, which is characteristic of all coherent imaging systems. In this process, correlated measurements are averaged (Anfinsen, Doulgeris, and Eltoft 2009). The multi-looking technique consists of first dividing and then separately processing N overlapped portions of the SAR bandwidth. The incoherent average of the so-obtained N SAR images improves the radiometric resolution by a factor of N. Its effective value can be quantified in terms of an equivalent number $N' \leq N$ of uncorrelated samples; this number is usually referred to as the ENL (Franceschetti and Lanari 1999).

In this respect, the ENL represents the number of independent intensity values averaged per pixel. It is often applied not just to describe the original data, but also to characterize the smoothing effect of post-processing operations such as image filtering (Oliver and Quegan 2004). In this case, ENL's purpose is to reflect grey-level variations induced by speckle. Variations caused by topography or ground cover must be excluded. Therefore, ENL has to be calculated over a homogeneous area within the SAR scene. For this study, an area fulfilling this criterion was selected manually. In total, the area covered by the selected pixels adds up to approximately 1200 km^2 or 230,000 pixels, and it is included by all used SAR scenes.

Results

In this section, the study results are presented. The following main steps can be distinguished: the computation of the benchmark image (temporal mean backscatter), the motivation for its application as a benchmark for the SAR image usability assessment, and its validation. Furthermore, the usability assessment of single-SAR imagery is covered.

Computation of the mean backscatter image

As an overview, the temporal mean backscatter, including the automatically derived permanent waterbody map, is presented in Figure 5. The mean backscatter was used as a benchmark, i.e. the optimum radar backscatter image, which is the basis for the usability

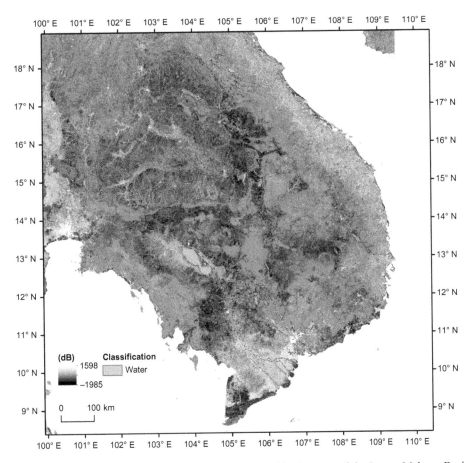

Figure 5.   This figure shows a map of the mean temporal backscatter of the Lower Mekong Basin (covering Cambodia, Vietnam, Laos, and Thailand). In blue, permanent waterbodies are overlaid.

assessment of single-SAR scenes. One by-product of this was a permanent waterbody classification. As mentioned above, the averaging period, and therefore the number of SAR scenes, influences the accuracy of the permanent waterbody map in two ways: first, by averaging over a higher number of scenes, image noise can be reduced, which leads to a lower number of misclassifications; second, a single-SAR scene represents the water coverage as a snapshot in time. By averaging over a certain time period, seasonal fluctuations were compensated and, therefore, the representation of the permanent waterbodies was improved. Figure 6 shows a close-up of the radar backscatter and classification results for a subset of the LMB. The images are based on a single-SAR scene, the average backscatter for one month, one year, and the entire data archive, respectively.

Suitability of mean backscatter image as a reference image

The quality of the mean backscatter images can be assessed in two ways. First, based on image quality, i.e. image noise and, second, based on the suitability for automatic waterbody classification. As a measure for image quality and level of noise ENL (Section 3.3) was used. Figure 7(a) shows the ENL dependence of the average number of images per

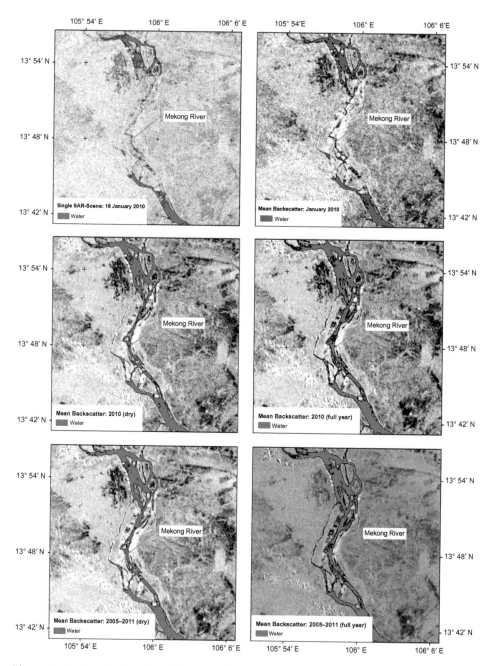

Figure 6. Backscatter for subset of the area of interest. Top-left: SAR scene as on 18 February 2010. Top-right: mean backscatter on January 2010. Centre-left: mean backscatter dry season during 2010. Centre-right: mean backscatter during 2010, dry season and wet season. Bottom-left: mean backscatter dry seasons 2005–2011. Bottom-right: mean backscatter 2005–2011, dry season and wet season.

pixel, averaged for the mean backscatter. The figure clearly shows that ENL increases steadily with a growing number of images, indicating lower levels of noise and therefore higher image quality.

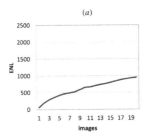
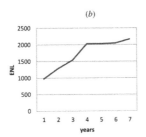
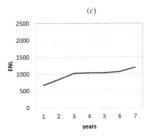

Figure 7. Relation between ENL and the number of averaged SAR images. The figures show the ENL in relation to the number of used images and the time frame. (*a*) Equivalent number of looks (ENL) related to the average number of images per pixel; (*b*) ENL related to the number of years used for averaging, all months; (*c*) ENL related to the number of years used for averaging, dry season.

Figures 7(*b*) and (*c*) show the ENL based on different averaging periods and image counts. Whereas in Figure 7(*b*) all months were included, Figure 7(*c*) was derived using only dry-season months. Both figures show similar characteristics – owing to an uneven distribution of images per year, the ENL does not grow linearly. The maximum ENL based on all months is significantly higher than that based on the dry season only. Using the entire seven years of SAR time-series data, the calculated ENL was approximately 2160 compared to approximately 1200 using data acquired during dry seasons only.

For an assessment of the mean backscatter-derived waterbody classification accuracy, the USGS SRTM Water Body data set was used. As a measure for the agreement between classifications and references data sets, confusion matrices were computed. In addition to the total accuracy, the user's accuracy, and the producer's accuracy, the kappa coefficient (Carletta 1996) was derived. The kappa coefficient is a measure of the improvement of a classification result over a purely random class assignment.

Figure 8 shows the waterbody classification accuracies dependent on the time periods used to derive the mean backscatter. All given accuracy measures are related to the water class. The graphics generally show a very high agreement between the SAR-based classification and the SRTM waterbodies. Despite the different levels of noise (Figure 7), the level of accuracy achieved for both approaches (using all months vs. using months belonging only to the dry season) is very similar, and both have a maximum kappa coefficient of about 0.9. The minimum size of waterbodies included in the SRTM waterbody data set is larger than the spatial resolution of the ASAR WS product. The SAR-derived classification therefore provides a higher level of detail (Figure 9), with a resulting negative impact on the user's accuracy.

The amount of data required is significantly less if only the dry season is used. The relation between ENL and kappa coefficient is shown in Figure 10; it shows that beyond a certain threshold value, no further accuracy improvement was achieved. This threshold was at about 1000, which corresponds to an averaging period of one year, approximately.

According to the above findings, the best results were achieved using the temporal mean backscatter, which was derived from all available SAR data. Therefore, this was used as a benchmark for the single-SAR image evaluation (Section 4.3). However, similar results could be achieved with a smaller SAR archive.

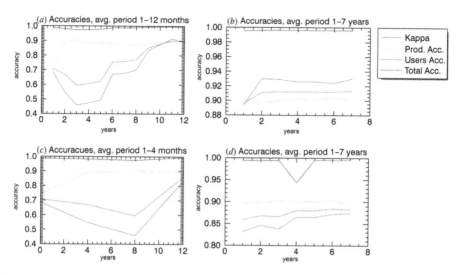

Figure 8. Accuracies of permanent waterbody classification in relation to the averaging time period. (*a*) Accuracies, average period 1–12 months; (*b*) accuracies, average period 1–7 years; (*c*) accuracies, average period 1–4 months; and (*d*) accuracies, average period 1–7 years.

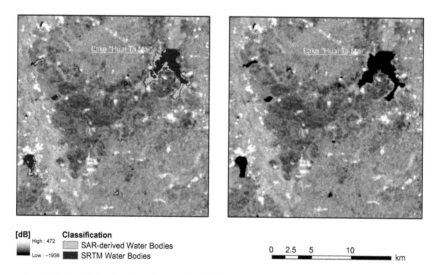

Figure 9. Left: SAR-backscatter image including the SAR-derived and the SRTM permanent waterbody mask. Right: optical satellite image (imagery ©TerraMetric) showing the waterbody extent.

Usability assessment of single-SAR images

Based on the location of high-contrast tiles, single-SAR scenes were evaluated on their suitability for automatic waterbody classification. Figure 11 shows a subset of the mean backscatter, which was derived from the SAR time-series database. The subset shows the area to the north and to the west of Lake Tonle Sap. In the left part of the image,

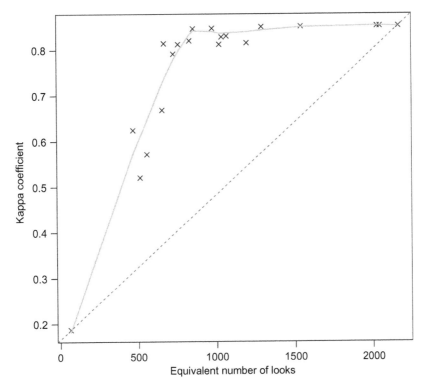

Figure 10. Dependency between the ENL (*x*-axis) and the classification accuracy, which is represented by the kappa coefficient (*y*-axis) (ENL vs. kappa coefficient).

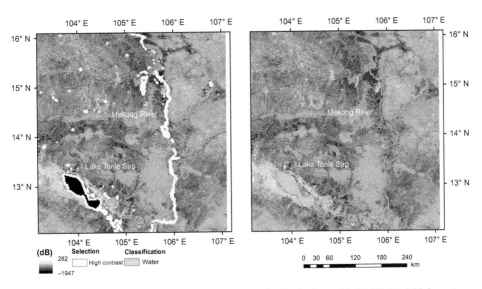

Figure 11. Mean backscatter for a subset around Lake Tonle Sap with highlighted high-contrast tiles (left) and resulting waterbody classification (right).

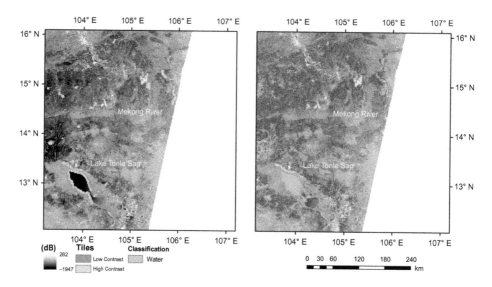

Figure 12. Result of contrast quantification, based on a single-SAR image, (left) and automatic waterbody selection (right).

high-contrast areas are highlighted in yellow. These are the automatically detected high-contrast tiles. It can be seen that the algorithm correctly identified tiles on the border of Lake Tonle Sap and along the Mekong River.

As mentioned above, the assumption was that the mean backscatter image can be used as a benchmark for identifying a SAR scene suitable for flood mapping. In the previous section it was shown that the derived permanent waterbody map corresponds very well to the SRTM waterbody data set, while providing a higher level of spatial detail. Justified by these results, the automatically derived permanent waterbodies were used for the quality assessment of the single-SAR-image-based classification results.

Figure 12 (left) shows the outcome of the evaluation process of a single-SAR scene acquired on 31 March 2011. Image tiles satisfying both selection parameters are coloured green; all the others are shown in red. In this example only a few tiles fulfil the requirements. The effect on the resulting waterbody classification is shown in Figure 12 (right) – in this case the algorithm could not retrieve an appropriate grey value separating water and no-water areas. This shows that even if a visual interpretation is possible, the automatic classification might fail. The percentage of tiles satisfying the contrast require-ments is approximately 22%, and the resulting waterbody classification has a kappa coefficient of 0.330.

Figures 13 and 14 show the results for another subset, which is located around Lake Nam Ngum in the northern part of the Mekong Basin, and was acquired on 18 June 2011. Figure 12 presents the mean backscatter and the location of high-contrast tiles within the subset. In this case approximately 60% of the selected tiles satisfy the selection criteria. Although most of the valid tiles are around Lake Nam Ngum, the accuracy corresponds to this trend. The kappa coefficient for this subset is 0.79.

Figure 14 shows the outcome of the data evaluation for the given image. This figure shows the dependency between the number of valid tiles and the classification accuracy. Figure 15 indicates a strong dependency between the percentage of high-contrast tiles and the classification accuracy.

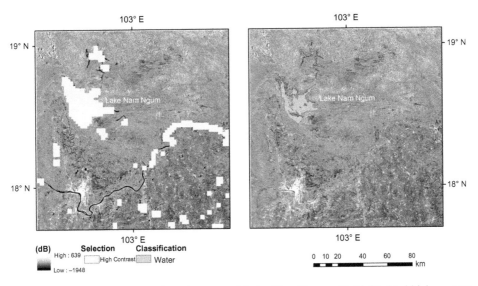

Figure 13.   Mean backscatter of a subset around Lake Nam Ngum with highlighted high-contrast tiles (left) and resulting waterbody classification (right).

Discussion

The relation between the surface-water classification accuracy (kappa coefficient) and the usability measure is shown in Figure 15. The two parameters exhibit a strong correlation with a correlation coefficient of 0.92. This demonstrates that the usability measure provides a reliable indication of the suitability of individual SAR scenes for flood mapping.

The benchmark used for classification quality assessment was based on an ASAR WS data archive. In general it can be said that by averaging over a certain time period, the

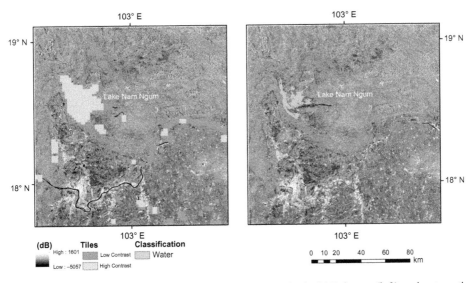

Figure 14.   Result of contrast quantification, based on a single-SAR image (left) and automatic waterbody selection (right).

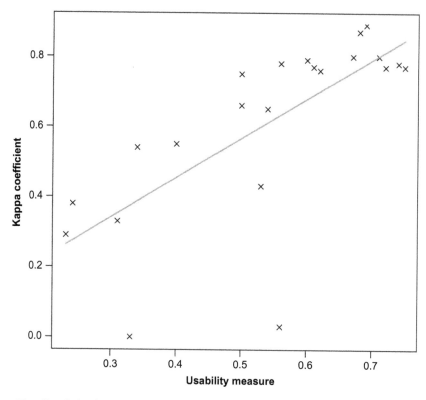

Figure 15. Correlation between usability measure and kappa coefficient (classification accuracies).

signal-to-noise ratio of SAR data can be improved. However, it was found that this is not necessarily reflected in the waterbody classification. Figure 7 shows, based on the year 2010, that the inclusion of the months February and March causes a decrease in quality. A probable explanation for this phenomenon is the beginning of the rainy season during these months – see Section 2 and Figure 2. It was found that, considering ASAR WS temporal resolution for LMB, the minimum averaging period should be about one year. Increasing the averaging period does not significantly improve the results. For other regions of the world or through application of other data sources, a shorter averaging period might be sufficient. Owing to the very high seasonal variability of the hydrological conditions in the LMB, averaging over a longer period was necessary.

However, some limitations were found during development. First, in the case of large differences between the minimum and the maximum extent of a waterbody, the pre-computed high-contrast tiles could be covered completely by water or land in a given SAR scene. Consequently, the tiles would not show sufficient contrast and could therefore falsely reduce the predicted usability measure. This might be the case around Lake Tonle Sap, which can show particularly large differences in extent between the dry and wet seasons. This can be recognized by visually assessing the tile selection map shown in Figures 12 and 14. Another issue is connected to waterbodies larger than approximately twice the size of the window used to identify high-contrast tiles. Rippling of the water surface due to wind or rain can complicate the water classification. If such deteriorating conditions occur at the centre of large waterbodies they will not be captured by the usability measure as tiles containing only water will not be identified as high-contrast tiles. This issue can be mitigated by increasing the tile size and it is not expected to represent a significant hurdle for flood mapping.

Conclusion

This article introduced a method for the quantification of the suitability of single-SAR images for automatic waterbody classification, the so-called usability measure. It was shown that the concept of high-contrast tiles could be used to compute the usability measure whereby unsuitable scenes could be automatically identified with an accuracy of more than 90%. The location of potential high-contrast tiles was identified based on a permanent water-body map derived from mean temporal backscatter. Furthermore, the approach provided an assessment of contrast variations within a single-SAR image.

Automatic pre-screening of SAR imagery is valuable in automatic processing environments and operational settings requiring rapid data delivery. Sentinel-1, for instance, is expected to provide a number of such services (European Space Agency 2013). The usability measure could also support manual SAR image interpretation. Another possible use of the proposed evaluation algorithm is automatic quality control in order to identify data containing artefacts, e.g. resulting from data transmission or processing errors. This is of particular interest considering the huge data amounts expected by future SAR missions.

Many applications in the field of remote sensing require detailed information about permanent waterbodies, e.g. in support of land-cover classification, improvement of digital elevation models (Slater et al. 2006), hydrological and biogeochemical modelling, GIS database updating, retrieval of geophysical parameters such as soil moisture (Hornacek et al. 2012), as well as flood prediction or flood monitoring (Tholey, Clandillon, and De Fraipont 1997). It was found that, in the case of ASAR WS data, a time series covering at least one year is necessary to derive accurate permanent waterbodies in the LMB.

The next generation of SAR sensors such as Sentinel-1 will offer higher spatial and radiometric resolution and a better signal-to-noise ratio. Snoeij et al. (2010) reported that the radiometric resolution of the Interferometric Wide-Swath mode of Sentinel-1 will improve upon the quality of the ASAR WS data with a factor of three. Combining data from such a sensor with the approach described in this work could further increase the accuracy and reliability of waterbody mapping with SAR.

Funding

This study was supported by funding from the German Aerospace Center (Deutsches Zentrum für Luft und Raumfahrt – DLR) within the framework of the WISDOM project. Furthermore, funding was provided by Capgemini Consulting Austria AG and the European Space Agency ESA-ESTEC as a part of the 'Fully Automated Aqua Processing Service' (FAAPS) project.

References

Anfinsen, S. N., A. P. Doulgeris, and T. Eltoft. 2009. "Estimation of the Equivalent Number of Looks in Polarimetric Synthetic Aperture Radar Imagery." *IEEE Transactions on Geoscience and Remote Sensing* 47 (11): 3795–3809. doi:10.1109/TGRS.2009.2019269.

Bazi, Y., L. Bruzzone, and F. Melgani. 2005. "An Unsupervised Approach Based on the Generalized Gaussian Model to Automatic Change Detection in Multitemporal SAR Images." *IEEE Transactions on Geoscience and Remote Sensing* 43 (4): 874–887. doi:10.1109/TGRS.2004.842441.

Brakenridge, R. G., E. Anderson, S. V. Nghiem, S. Caquard, and T. B. Shabaneh. 2003. "Flood Warnings, Flood Disaster Assessments, and Flood Hazard Reduction: The Roles of Orbital Remote Sensing." Proceedings of the 30th international symposium on remote sensing of environment. Pasadena, CA : Jet Propulsion Laboratory, National Aeronautics and Space Administration.

Carletta, J. 1996. "Squibs and Discussions Assessing Agreement on Classification Tasks: The Kappa Statistic." *Computational Linguistics* 22 (2): 249–254.

Dobson, M. C., L. E. Pierce, and F. T. Ulaby. 1996. "Knowledge-Based Land-Cover Classification Using ERS-1/JERS-1 SAR Composites." *IEEE Transactions on Geoscience and Remote Sensing* 34 (1): 83–99. doi:10.1109/36.481896.

European Space Agency. 2013. *Sentinel-1 User Handbook.* Paris: European Space Agency.

Franceschetti, G., and R. Lanari. 1999. *Synthetic Aperture Radar Processing.* Boca Raton, FL: CRC Press.

Gagnon, L., and A. Jouan. 1997. "Speckle Filtering of SAR Images: A Comparative Study between Complex-Wavelet-Based and Standard Filters." In *SPIE 3169, Wavelet Applications in Signal and Image Processing*, edited by A. Aldroubi, A. F. Laine, and M. A. Unser, 80–91.

Hoanh, C. T., H. Guttman, P. Droogers, and J. Aerts. 2003. *Water, Climate, Food, and Environment in the Mekong Basin in Southeast Asia.* Report by International Water Management Institute (IWMI), the Mekong River Commission Secretariat (MRCS), and the Institute of Environmental Studies (IVM).

Hornacek, M., W. Wagner, D. Sabel, H.-L. Truong, P. Snoeij, T. Hahmann, E. Diedrich, and M. Doubkova. 2012. "Potential for High Resolution Systematic Global Surface Soil Moisture Retrieval via Change Detection Using Sentinel-1." *IEEE Journal of Selected Topic in Applied Earth Observations and Remote Sensing* 5 (2): 1303–1311.

Hostache, R., P. Matgen, and W. Wagner. 2012. "Change Detection Approaches for Flood Extent Mapping: How to Select the Most Adequate Reference Image from Online Archives?" *International Journal of Applied Earth Observation and Geoinformation* 19: 205–213. doi:10.1016/j.jag.2012.05.003.

Juren, G., and M. I. Khan. 2010. *Pakistan Floods 2010.* Islamabad: Asian Development Bank.

Kuenzer, C., H. Guo, J. Huth, P. Leinenkugel, X. Li, and S. Dech. 2013. "Flood Mapping and Flood Dynamics of the Mekong Delta: Envisat-ASAR-WSM Based Time Series Analyses." *Remote Sensing* 5 (2): 687–715.

Martinez, J., and T. Letoan. 2007. "Mapping of Flood Dynamics and Spatial Distribution of Vegetation in the Amazon Floodplain Using Multitemporal SAR Data." *Remote Sensing of Environment* 108 (3): 209–223. doi:10.1016/j.rse.2006.11.012.

Martinis, S., A. Twele, and S. Voigt. 2009. "Towards Operational Near Real-Time Flood Detection Using a Split-Based Automatic Thresholding Procedure on High Resolution TerraSAR-x Data." *Natural Hazards and Earth System Science* 9 (2): 303–314. doi:10.5194/nhess-9-303-2009.

Naeimi, V., P. Leinenkugel, D. Sabel, W. Wagner, H. Apel, and C. Kuenzer. 2013. "Evaluation of Soil Moisture Retrieval from the ERS and Metop Scatterometers in the Lower Mekong Basin." *Remote Sensing* 5 (4): 1603–1623. doi:10.3390/rs5041603.

Oliver, C., and S. Quegan. 2004. *Understanding Synthetic Aperture Radar Images.* Raleigh, NC: SciTech.

Otsu, N. 1979. "A Threshold Selection Method from Gray-Level Histograms." *IEEE Transactions on Systems, Man, and Cybernetics* 9 (1): 62–66. doi:10.1109/TSMC.1979.4310076.

Pathe, C., W. Wagner, D. Sabel, M. Doubkova, and J. B. Basara. 2009. "Using Envisat ASAR Global Mode Data for Surface Soil Moisture Retrieval over Oklahoma, USA." *IEEE Transactions on Geoscience and Remote Sensing* 47 (2): 468–480. doi:10.1109/TGRS.2008.2004711.

Slater, J. A., G. Garvey, C. Johnston, J. Haase, B. Heady, G. Kroenung, and J. Little. 2006. "The SRTM Data "Finishing" Process and Products." *Photogrammetric Engineering Remote Sensing* 72 (3): 237–247. doi:10.14358/PERS.72.3.237.

Snoeij, P., E. Attema, M. Davidson, B. Duesmann, N. Floury, G. Levrini, B. Rommen, and B. Rosich. 2010. "Sentinel-1 Radar Mission: Status and Performance." *IEEE Aerospace and Electronic Systems Magazine* 25 (8): 32–39. doi:10.1109/MAES.2010.5552610.

Tholey, N., S. Clandillon, and P. De Fraipont. 1997. "The Contribution of Spaceborne SAR and Optical Data in Monitoring Flood Events: Examples in Northern and Southern France." *Hydrological Processes* 11 (10): 1409–1413. doi:10.1002/(SICI)1099-1085(199708) 11:10<1409::AID-HYP531>3.0.CO;2-V.

United Nations International Strategy for Disaster Reduction (UNISDR). 2011. *Disasters in Numbers.* Geneva: UNISDR.

Woodhouse, I. H. 2005. *Introduction to Microwave Remote Sensing.* 1st ed. Boca Raton, FL: CRC Press.

Ecosystem assessment in the Tonle Sap Lake region of Cambodia using RADARSAT-2 Wide Fine-mode SAR data

Lu Zhang, Huadong Guo, Xinwu Li, and Liyan Wang

Key Laboratory of Digital Earth Science, Institute of Remote Sensing and Digital Earth, Chinese Academy of Sciences, Beijing, China

The Tonle Sap Lake (TSL), located in Cambodia, is the largest freshwater lake in Southeast Asia and has significant ecological, economic, and sociocultural value. The TSL's ecosystems have been affected by climate change and an increasing amount of human activity in recent years. Considering that the TSL area is often covered by clouds, particularly in the rainy season, synthetic aperture radar (SAR) data are suitable for assessing the ecosystems in this great lake, as SAR enables weather- and cloud-independent observations. In this study, we investigated the capability of the RADARSAT-2 Wide Fine (WF) mode dual-polarization SAR data with a scene size of 170×150 km (azimuth × range) and a resolution of 7.6×5.2 m to study TSL's ecosystem, by analysing the usefulness of backscattering coefficients and scattering mechanism-related parameters in identifying artificial targets and different land-cover types. The results of this study demonstrate the applicability of RADARSAT-2 WF-mode SAR data in the study of TSL's ecosystems.

1. Introduction

The Tonle Sap Lake (TSL), located in Cambodia, is a natural reservoir of the Mekong River and is the largest freshwater lake in Southeast Asia. Increasing attention has been paid to the large lake and its river system because of their significant ecological, economic, and socio-cultural values (Penny 2006). Recently, human activities and global environmental changes have significantly influenced the ecological environment of TSL (Frappart et al. 2006). The challenges facing the basin include the hydropower-induced changes in flood pulse, annual and occasionally severe flooding events, and water resource distribution in general (Kuenzer 2013; Kuenzer, Campbell, et al. 2013; Naeimi et al. 2013). Establishing the means to monitor land use and the ecological environment changes efficiently and analysing the mechanisms of these changes are of international interest for researchers (Sarkkula and Koponen 2003; Sithirith 2006; Van Trung, Choi, and Won 2009; Kuenzer, Guo, Huth, et al. 2013; Kuenzer, Guo, Schlegel, et al. 2013).

Synthetic aperture radar (SAR) plays an important role in assessing the ecosystems in the TSL area because of its ability to provide weather- and cloud-independent observations and its sensitivity to the vegetation structure (Guo 2000; Brisco et al. 2008; Leinenkugel, Kuenzer, and Dech 2013; Leinenkugel, Kuenzer, Oppelt, ert al. 2013; Kuenzer, Guo, Huth, et al. 2013). Many studies on wetlands have been conducted using multi-band and multi-scale SAR data. Van Trung, Choi, and Won (2009, 2010, 2011) studied the land-cover changes of the TSL region using ALOS PALSAR L-band SAR data. Choi, Lee, and Won (2009) monitored and measured the variation of the water level in the flood area of the TSL based on the SAR interferometry technique, which can be used to generate maps of surface deformation or digital elevation from differences in the phase of two or more SAR sets of data. Milne and Tapley (2004) carried out some studies of wetland ecosystems in 2000 for the TSL basin and demonstrated the capability of multi-band and multi-polarization AIRSAR data in the study of mapping the wetland vegetation. Recently, Kuenzer, Guo, Huth, et al. (2013) and Kuenzer, Guo, Schlegel, et al. (2013) studied the flood dynamics and situation of the Mekong Delta from 2007 to 2011 using multi-scale SAR data acquired by Envisat-ASAR and TerraSAR-X. The studies mentioned above demonstrate the applicability of multi-parameter SAR data in the study of TSL's ecosystem, especially to the studies of flood situation and land-cover mapping.

Recently, high-resolution and wide-swath SAR data have become one of the important development directions of SAR techniques of the future (Gabele and Krieger 2008; Guo and Li 2011). These types of data can overcome the challenge of collecting high-resolution and large-area SAR data simultaneously and are very useful in large-area and fast-changing study areas, such as the TSL region. RADARSAT-2 Wide Fine (WF)-mode dual-polarization SAR data with a scene size of 170 × 150 km (azimuth × range) and a resolution of 7.6 × 5.2 m was used to assess the ecosystem of TSL. By investigating the capability of the RADARSAT-2 WF-mode SAR data on artificial target detection and land-cover classification, this study aims to (1) assess the applicability of the RADARSAT-2 WF-mode SAR data to study wetland ecosystems, and (2) to gain a better understanding of the characteristics of the TSL wetland ecosystem.

2. Study area and SAR data

2.1. Tonle Sap Lake

TSL was chosen as the study area for this study. Figure 1 shows the geographic location of the study area. The images in Figure 1 were obtained from Landsat-8 optical images

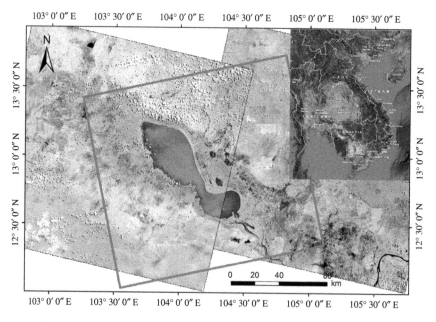

Figure 1.   The location of the TSL (Landsat-8 data acquired on 30 May and 9 June 2013). The red rectangle indicates the footprint of RADARSAT-2 WF SAR data acquired for TSL.

acquired on 30 May and 9 June 2013, respectively. During the dry season, water flows into the Mekong River from the TSL. The lake is usually 150 km in length and 30 km in width during the dry reason. The surface area of the lake is approximately 2700–3000 km^2, and the average water depth is 1 m. In contrast to the dry season, the water flows from the Mekong River via the TSL river into the TSL during the wet season (July–October), reducing flooding downstream in the Mekong River basin. In the wet season, the surface area of the TSL expands to more than 10,000 km^2, the width of the lake increases to 100 km, and the average water depth increases to over 12 m. Consequently, the TSL plays an important role in mitigating extremes of seasonal hydrology associated with the contrasting wet and dry seasons because of its unique feature of reverse flow (Frappart et al. 2006).

In the areas surrounding the TSL, 1.7 million people live in the 1555 villages scattered alongside the lake (as of March 2008) (Marko et al. 2011). Most of them make a living from the abundant fisheries. Some villages are entirely floating villages perfectly adapted to the rise and fall of water levels. There are many floating or stilted buildings in this region, as well as boats and fishery equipment. The distribution of these targets can be used as a descriptor of human activity to estimate the effect of this activity on the TSL's ecosystem. Figure 2 shows pictures of some of these artificial targets in the TSL area.

2.2. Comparison of remote-sensing data for the TSL area

Many types of remote-sensing data for the TSL study area can be obtained. The types of data available vary in their resolution and swatch width. Table 1 lists several parameters of some commonly used remote-sensing data, including Landsat series optical data, RADARSAT-2 WF- and WFQ-mode SAR data, Envisat-ASAR WS- and AP-mode

Figure 2. Pictures of artificial targets in TSL, including flooded or stilted buildings and docked boats.

Table 1. Comparison of some remote-sensing data used in the TSL area.

Sensors	Mode	Band	Swath width (km)	Resolution (m) (Range × Azimuth)	Polarization
Landsat		Optical	185	30	
RADARSAT-2	Wide Fine	C	150	5.2 × 7.7	Single or Dual*
RADARSAT-2	WFQ*	C	50	5.2 × 7.6	Quad*
Envisat-ASAR	WS*	C	400	150	Single Co*.
Envisat-ASAR	AP*	C	100	30	Dual
ALOS-PALSAR	ScanSAR	L	250–350	100	Single Co.
ALOS-PALSAR	High res. (pol)	L	20–65	20	Quad
TerraSAR-X	Stripmap	X	15 (Single)	3.5 × 3.5	Single Co. or Dual
			30 (dual)	6.6 × 6.6	
TerraSAR-X	ScanSAR	X	100	19	Single Co.

Note: *WFQ, wide fine quad mode; WS, wide swath mode (ScanSAR); AP, alternating polarization; Single, HH or VV or HV or VH; Single Co., HH or VV; Quad, HH/HV/VH/VV; Dual, HH/VV or HH/HV or VV/VH (HH/HV or VV/VH only for PALSAR data).

SAR data, ALOS-PALSAR ScanSAR- and HR-mode SAR data, and TerraSAR-X StripMap- and ScanSAR-mode SAR data.

As Table 1 shows, Landsat data are excellent for long-term and large-scale study because of their wide swath, medium spatial resolution, and high temporal resolution. However, Landsat data are not well suited to analyse areas with frequent cloud cover. Envisat-ASAR WS mode and ALOS-PALSAR ScanSAR mode can acquire SAR data with very large swath widths of over 250 km. These data are suitable for large-scale study, such as general flood patterns in this area. However, their resolutions are too low to meet some of the study requirements such as artificial target detection for the TSL region. RADARSAT-2 WFQ data, Envisat-ASAR AP data, ALOS-PALSAR polarimetric data, and TerraSAR-X data, with their high- or medium-resolution and multi-channel information, can be useful in high-resolution land-cover mapping and urban flood assessment in a small-area region, but their swath widths are so narrow that they cannot acquire a single dataset that covers all or most of the TSL region. The RADARSAT-2 WF data used in this study, with a resolution of 7.6 × 5.2 m (azimuth × range), a swath width of 150 km, and dual-polarization information, are well suited for the studies of artificial target detection and high-resolution land-cover classification in the TSL area.

2.3. RADARSAT-2 WF data for the TSL and preprocessing

RADARSAT-2 C-band dual-polarization (HH/HV) WF-mode SAR data acquired on 20 December 2012 were used in this study. The incidence angle of the RADARSAT-2 data ranges from 19° to 31°. The resolutions in the azimuth and range directions are 7.7 m and 5.2 m, respectively, and the swath width is 150 km. The footprint of the data is indicated in Figure 1 by a red rectangle.

The covariance matrix (the C_2 matrix) extracted from the RADARSAT-2 WF-mode dual-polarization data is shown in Equation (1). C_{11} and C_{22} are the backscattering coefficients of the HH and HV channels, respectively. C_{12} and C_{21} are the correlation coefficients between the HH and HV channels and between the HV and HH channels, respectively. $C_{12} = C_{21}$*, with * the conjugate operator and the two are equal in amplitude. Figure 3 is the red–green–blue (RGB) pseudo-colour composite image of the

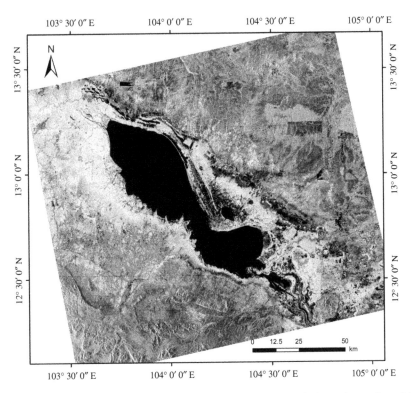

Figure 3. RGB pseudo-colour image of the RADARSAT-2 WF-mode SAR data collected for the TSL area. (Red: C_{11} (HH); Green: C_{22} (HV); Blue: C_{12} (HH·HV*)).

RADARSAT-2 SAR data for the TSL area, where the red, green, and blue colours represent the amplitudes of C_{11}, C_{22}, and C_{12}, respectively.

$$\mathbf{C}_2 = \begin{bmatrix} C_{11} & C_{12} \\ C_{21} & C_{22} \end{bmatrix} = \begin{bmatrix} \langle S_{HH}S_{HH}^* \rangle & \langle S_{HH}S_{HH}^* \rangle \\ \langle S_{HV}S_{HH}^* \rangle & \langle S_{HV}S_{HH}^* \rangle \end{bmatrix}. \tag{1}$$

Before obtaining the TSL's ecosystem-related information from the SAR data, SAR image preprocessing, such as calibration, filtering, geocoding, and incidence angle compensation, was necessary. In this study, a box car filter with a window of 5 × 5 was applied to the covariance matrix ($\mathbf{C}_2$ matrix) to suppress speckle noise in the SAR data. In addition, considering that the range of the incidence angle of the RADARSAT-2 FW data is relatively large (12°) compared with traditional StripMap data, Lambert's law was applied to compensate globally for the effect of the incident angle (Ardila, Tolpekin, and Bijker 2010). In Equation (2), σ^0 is the calibrated backscattering coefficient using a sigma lookup table file, θ is the incidence angle, and σ^0_m is the backscattering coefficient after incidence angle compensation. In this study, the real and imaginary components of the HH and HV channels, respectively, were compensated using the square root of the cosine in Equation (2).

$$\sigma^0 = \sigma^0_m \cos\theta. \tag{2}$$

This compensation method cannot completely compensate for the effect of the incidence angle because the change trends with incidence angle vary in different land-cover types, but it can be used to make the SAR data appear more balanced in the range direction.

3. Information extraction methods applied to SAR data

The significant ecosystem-related information for the TSL that can be extracted from RADARSAT-2 WF SAR data mainly includes the distribution of artificial targets and the distribution of land-cover types. This section describes the characteristics of some SAR parameters that are sensitive to ecosystem-related information and discusses some methods for obtaining this information.

3.1. Artificial target extraction methods

SAR data are sensitive to artificial targets that can produce corner reflections. RADARSAT-2 C-band SAR data are especially suitable for detecting artificial targets on the water surface of TSL, such as boats, buildings, and fishery equipment, because of their characteristics of high resolution and shorter wavelength, which have a higher sensitivity to small scatterers, compared with the sensitivity of longer-wavelength SAR data (Lee and Pottier 2009).

Many target detection methods for SAR data have been proposed (Ferro-Famil et al. 2003, Ferro-Famil, Reigber, and Pottier 2005; Zhang, Guo, and Han 2007; Migliaccio et al. 2011). Some of these methods are based on the fact that artificial targets have higher backscattering coefficients than the background. These methods are effective in the analyses of SAR data for the TSL area because the backscattering coefficients of the wetland and lake are obviously lower than those of artificial targets. Some methods, such as the HH/HV correlation coefficient method (HV_c method) (Migliaccio et al. 2011), are based on the reflection symmetry characteristics of the observed targets. The theoretical basis of the HV_c method is that natural targets are often regarded as reflection symmetry targets, whereas artificial targets do not satisfy reflection symmetry because of the complex shapes associated with their plane, dihedral, and trihedral structures. The correlation coefficient between the like- and cross-polarized scattering amplitudes, or HV_c, shown in Equation (3), can be used as a measure of the reflection symmetry. S_{HH} and S_{HV} in Equation (3) are the backscattering values of the HH and HV polarized channels, respectively, and the notation $\| \ \|$ indicates the absolute value.

$$HV_c = \left\| S_{HH} S_{HV}^* \right\|.$$ (3)

Figure 4 shows a comparison of the HH and HV backscattering coefficients and the HV_c parameter between the artificial targets and their background, such as waterbodies and vegetation. More than 1000 pixels were chosen for each type, and the DB values of the pixels were used to show the differences clearly.

As Figure 4 shows, the HH backscattering coefficient is the most effective differentiator of artificial targets from their background, including both waterbodies and vegetation. The HV backscattering coefficient and the HV_c parameter can differentiate between artificial targets and waterbodies; however, they are not as effective as the HH backscattering coefficient in distinguishing between artificial targets and vegetation. The HV

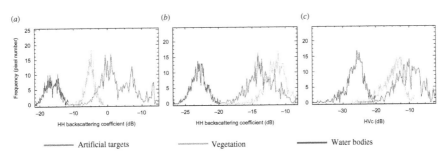

Figure 4. Comparisons of the (*a*) HH and (*b*) VV backscattering coefficients and (*c*) the HV_c parameter for artificial targets, waterbodies, and vegetation.

backscattering coefficient of the artificial targets is lower than that of vegetation because of its sensitivity to volume scattering usually caused by vegetation. Because docked boats and houses are often surrounded by vegetation in the TSL area, the HH backscattering coefficient was used in this study to detect the artificial targets, and a simple threshold value (0 dB) was artificially set, based on Figure 4(*a*). The artificial target detection step is shown on the left-hand side of the flow chart in Figure 5.

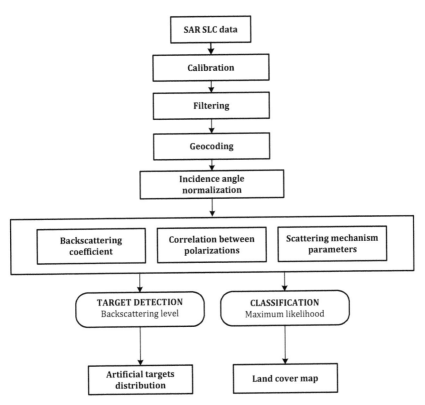

Figure 5. Flow chart of the process of extraction of the wetland's ecosystem-related information from RADARSAT-2 SAR data.

3.2. Land-cover classification

The classification technique belongs to the most important and commonly used methods for deriving land-cover and land-use distributions, which is important to obtain information for research on the ecosystem of the TSL area. Many parameters derived from RADARSAT-2 WF-mode dual-polarization data can be used to describe the features of land-cover types, including not only the backscatter coefficient with different polarizations but also the mechanism-related parameters (scattering entropy, scattering type angle, anisotropy, etc.), which can provide valuable information about an observed target's structure and are very useful in characterizing flooded areas (Brisco et al. 2013; Kuenzer, Guo, Huth, et al. 2013; Kuenzer, Guo, Schlegel, et al. 2013).

The backscattering coefficients with HH and HV polarization can be directly obtained from the SAR data. The values of the scattering mechanism-related parameters were derived using the target matrix decomposition technique. The $H/A/\alpha$ model (Cloude and Pottier 1996, 1997) is commonly used to extract the values of scattering mechanism-related parameters, such as the scattering type angle (α), the coherency entropy (H), and the anisotropy (A), from polarimetric SAR data. The $H2\alpha$ decomposition model (Cloude 2007) is a modification of the $H/A/\alpha$ model that was developed for use with dual-polarization SAR data. The values of the dual-polarization entropy (Dual-H), dual-polarization anisotropy (Dual-A), and dual-polarization scattering type angle (Dual-α) parameters can be obtained using the $H2\alpha$ model. These parameters are useful in wetland research, although only part of the scattering-type information can be obtained from dual-polarization SAR data.

In the TSL region, the spatial distributions and changes of several land-cover types, including waterbodies, settlements, forest, agriculture, grass, flooded forest, and flooded shrubs, are important for ecological study. Figure 6 shows the histograms of the above-mentioned SAR parameters for these land-cover types, and the mean values and standard deviation values are listed in Table 2. To facilitate demonstration of the characteristics of the scattering mechanism-related parameters, an entropy-scattering-type angle plane (H–α plane) (Cloude 2007) was used for some land-cover types and is shown in Figure 7.

As Figures 6 and 7, and Table 2 show, the distribution characteristics of both HH and HV backscattering coefficients and the Dual-H, Dual-A, and Dual-α scattering mechanism-related parameters differ for different typical land-cover types in the study area. These different parameters have their respective advantages in differentiating among land-cover types in the TSL area. For example, waterbodies can be clearly distinguished from the backscattering coefficients parameters by its lower value compared with the other land-cover types, and bare soil can be easily distinguished from the scattering mechanism-related parameters because of its scattering characteristic of simple plane-surface backscattering, with low values of entropy and scattering-type angle. The backscattering coefficients of forest and flooded forest are similar, but they can be distinguished from each other using the scattering-type angle alpha. As a result, the combination of HH and HV backscattering coefficients and scattering mechanism-related parameters Dual-H, Dual-A, and Dual-α can effectively differentiate among the land-cover types in the TSL area.

In addition, the values of these parameters are approximately normally distributed, and therefore the maximum likelihood classification method can be used to obtain land-cover information for the TSL area. The maximum likelihood classification method is a supervised classification method for calculating the probability that a given pixel belongs to a specific class, based on the assumption that the statistics for each class in each band

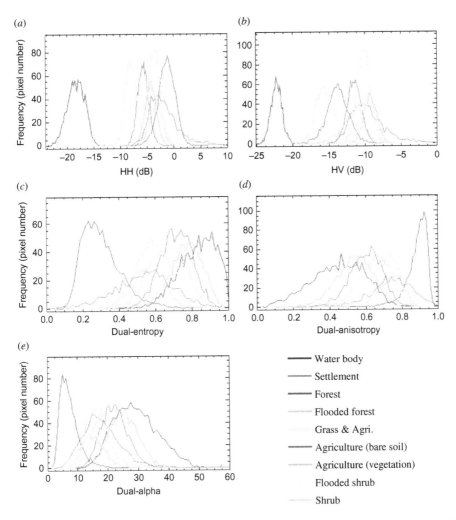

Figure 6. Histograms of the (*a*) HH and (*b*) HV backscattering coefficients, (*c*) entropy, (*d*) anisotropy, and (*e*) scattering-type angle of dual-polarization SAR data for different land-cover types.

are normally distributed (Mather and Tso 2010). The process for obtaining land-cover information appears on the right-hand side of the flow chart shown in Figure 5.

4. Results and discussion

4.1. *Artificial target detection of the TSL region*

Using the artificial target detection method based on the HH backscattering coefficient mentioned in Section 3.1, the distribution of the artificial targets was obtained and is shown in Figure 8. However, the artificial targets detected from the RADARSAT-2 WF-mode SAR data are too small to be visible in the entire image. Therefore, we used some detailed images to describe the artificial target detection results, indicated by the red colour. These detailed images are shown as (*a*)–(*g*) in Figure 8, and their locations are

Table 2. Mean values and standard deviation values of the backscattering coefficients and scattering mechanism-related parameters for different land-cover types in TSL.

Classes	Backscattering coefficients mean (std)		Scattering mechanism-related parameter mean (std)		
	C-HH (dB)	C-HV (dB)	Dual-entropy	Dual-alpha	Dual-anisotropy
Waterbody	−18.32(1.63)	−22.24(0.77)	0.83(0.10)	29.19(7.20)	0.45(0.15)
Forest	−5.71(1.06)	−11.52(1.00)	0.71(0.09)	22.15(4.41)	0.60(0.09)
Settlement	−1.67(3.12)	−9.54(2.55)	0.54(0.16)	18.05(6.44)	0.74(0.12)
Agriculture (bare soil)	−1.35(1.59)	−13.83(1.52)	0.31(0.11)	7.37(3.22)	0.89(0.06)
Grass and agriculture	−7.99(1.10)	−15.51(1.29)	0.60(0.09)	16.69(3.76)	0.71(0.07)
Agriculture (vegetation)	−4.31(1.04)	−9.28(0.93)	0.77(0.09)	25.19(4.91)	0.54(0.10)
Flooded shrub	−2.65(0.91)	−9.68(0.89)	0.63(0.09)	18.17(3.90)	0.68(0.07)
Shrub	−3.62(1.11)	−10.17(0.95)	0.66(0.10)	19.71(4.46)	0.65(0.09)
Flooded forest	−3.55(1.37)	−11.69(1.51)	0.55(0.09)	15.17(3.57)	0.74(0.07)

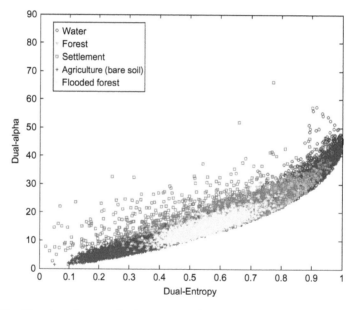

Figure 7. The H–α plane for some land-cover types in TSL.

marked by the blue rectangles in the upper left panoramic image. Based on the results, because there were no fishing operations under way at the time of the acquisition of the SAR data (20 December), no boat or fishery activity can be detected on the lake from the SAR data. However, many docked boats, floating buildings, and stilted buildings on the small rivers around the lake could be clearly detected. In summary, the artificial target distribution information can be used to assess the extent of human activities, particularly fishery activities. The artificial target distribution indicates that the effects of human activities in the north of the TSL area are more significant than those in the south of the TSL area.

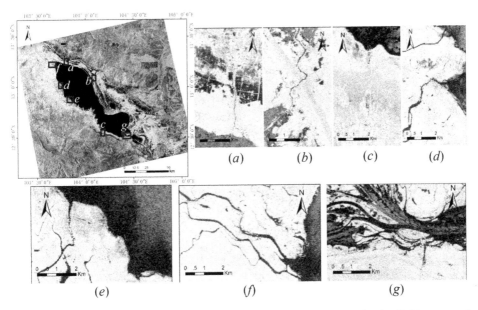

Figure 8. Artificial target detection results. The images from (*a*) to (*g*) are the detailed images, and the artificial targets detected are marked in red. The locations of these detailed images are marked by the blue rectangles in the upper-left panoramic image of the TSL.

The key characteristics of RADARSAT-2 WF-mode SAR data are its high resolution, wide swath, and short wavelength (C-band). These features make these data effective in detecting artificial targets in the TSL, such as small boats, fishery equipment, and floating buildings. The sensitivity of WF SAR data to small artificial targets is attributable to the key characteristics of high resolution and short wavelength. In addition, the characteristic of wide swath can make WF-mode data cover the entire study region at one time, thereby avoiding mistakes induced by the multi-temporal mosaic process, especially for moving artificial targets, such as boats, which are rarely stationary.

4.2. Land-cover map of the TSL region

Based on the method described in Section 3.2, a land-cover classification map can be obtained for the TSL area from RADARSAT-2 FW SAR data. The classes include some land-cover types that are typical in wetlands, such as flooded forest, flooded shrubs, grass, shrubs, forest, agriculture, bare soil, and settlement. Approximately 5000 pixels for each class were chosen as training samples for the maximum likelihood classifier, based on the high-resolution optical data of the TSL area and the land-cover classification results reported by Van Trung, Choi, and Won (2010). The classification results are shown in Figure 9. More than 1000 additional validation samples were artificially chosen from high-resolution Google Earth imagery for each class to verify the land-cover classification results. The confusion matrix is shown in Table 3. The overall accuracy and the kappa coefficient are 85% and 0.83, respectively. The results indicate that the typical land-cover types in the TSL area can be effectively segmented using RADARSAT-2 WF-mode SAR data. Based on the land-cover map derived from RADARSAT-2 data, the characteristics of some land-cover types are discussed as follows.

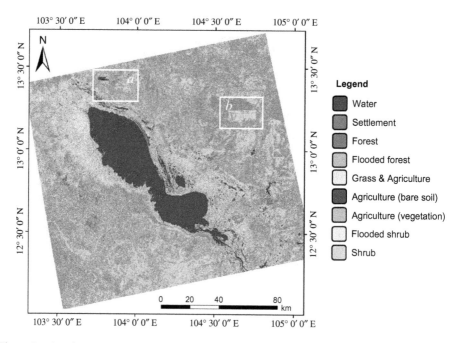

Figure 9. Land-cover map of the TSL region obtained from RADARSAT-2 WF-mode SAR data.

4.2.1. Forest and human settlement

As Figure 9 shows, the wetlands are mainly located in the periphery of the lake and river waterbodies. Nearby are settlements and agricultural areas. The forest is located in the outer region. Figure 10 shows a detailed land-cover map of the areas near Angkor Wat, a famous archaeological site located in the northeast of the TSL area and marked by the white rectangle labelled 'a' in Figure 9. The urban area marked in red and the agricultural land indicated by the brown and sallow colours can be observed clearly in Figure 10. The notable expansion of human settlements and agricultural land use in the TSL is caused by the growth in tourism in the Angkor Wat area.

Another detailed image of the land-cover map is shown in Figure 11, which indicates the severe deforestation in the northwest of the TSL area. The location is marked by the white rectangle labelled 'b' in Figure 9. By comparing these images with the Landsat TM image of this area acquired in 2009, shown in Figure 11(a), it is clear that almost half of the forest area has been logged. Figures 10 and 11 show the capability of C-band SAR data on the study of settlement expansion and deforestation. These results also indicate the effects of human activities on the TSL have been severe recently.

4.2.2. Wetland

For better obtaining the land-cover map of the TSL wetland, the outer area of the TSL was removed for decreasing the effects of non-wetland land-cover types. Figure 12 shows the results, mainly including waterbody, flooded forest, flooded shrub, flooded grass, upland forest, upland shrub, bare soil, and rice fields. As Figure 12 shows, the fine distribution of these land-cover types can be obtained. Flooded shrub and grass types are the two principal land-cover types in the TSL area. Flooded forest type is scattered around the

Table 3. Confusion matrix of the land-cover classification.

Class	Water	Forest	Settlement	Agriculture (bare soil)	Grass & agriculture	Flooded shrub	Shrub	Flooded forest	Agriculture (vegetation)	Total
Water	1180	0	0	0	0	9	0	0	0	1189
Forest	0	1113	0	0	9	0	0	13	0	1135
Settlement	0	0	1150	6	1	1	0	8	0	1166
Agriculture	0	0	149	1205	0	49	2	44	0	1449
Bare soil	79	45	0	0	1142	0	0	0	0	1266
Flooded shrub	0	0	1	18	0	1198	34	0	0	1251
Shrub	0	0	0	5	0	93	1174	7	340	1619
Flooded forest	97	157	58	103	173	31	59	1280	149	2107
Grass	0	0	0	0	0	2	49	12	812	875
Total	1356	1315	1358	1337	1325	1383	1318	1364	1301	12057
Producer's accuracy (%)	87.02	84.64	84.68	90.13	86.19	87.19	89.07	93.84	62.41	
User's accuracy (%)	99.20	98.06	98.63	83.16	90.21	95.76	72.51	60.75	92.80	
Overall accuracy (%)	85.05									
Kappa coefficient	0.8317									

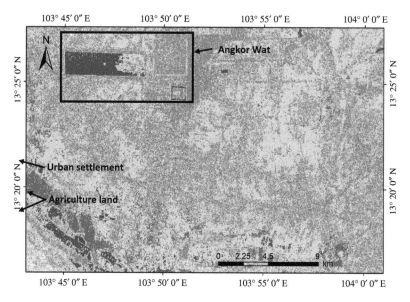

Figure 10. A detailed land-cover map of the areas near Angkor Wat.

flooded shrub and grass types, mainly concentrating on the edge of the TSL Lake and the river systems. In addition, according to the land-cover map, it is difficult to differentiate flooded types from upland types because the penetrating capability of short wavelength SAR data is weaker compared to that of long wavelength SAR data and it is difficult for electromagnetic waves to penetrate to the ground or the water surface. Therefore misclassification often occurs between them.

In summary, RADARSAT-2 WF-mode SAR data are ideal for use in studying land cover and land use in the TSL region. The high resolution and dual polarization of RADARSAT-2 WF-mode SAR data make it possible to obtain finely detailed land-cover maps of the TSL area. In addition, the formation of a mosaic of multi-temporal images is not necessary to use RADARSAT-2 WF-mode SAR data, which makes it possible to avoid introducing errors associated with using images acquired at different times and avoid the considerable amount of work required for the multi-temporal mosaic process.

5. Conclusions

The TSL, the largest freshwater lake in Southeast Asia, has significant ecological, economic, and sociocultural value. SAR remote-sensing techniques play an important role in assessing the ecosystems in the TSL area, especially because the effectiveness of such techniques is not constrained by weather conditions and cloud cover. RADARSAT-2 WF-mode SAR data acquired for the TSL area were used in this ecological study. The key characteristics of the RADARSAT-2 WF-mode SAR data are its high resolution (7.6 × 5.2 m), wide swath (170 × 150 km, dual-polarization (HH/HV), and short wavelength (C-band).

By focusing on the significant ecosystem-related information in the TSL, such as the distributions of artificial targets and land-cover types, values of the backscattering coefficients and scattering mechanism-related parameters were extracted from the RADARSAT-

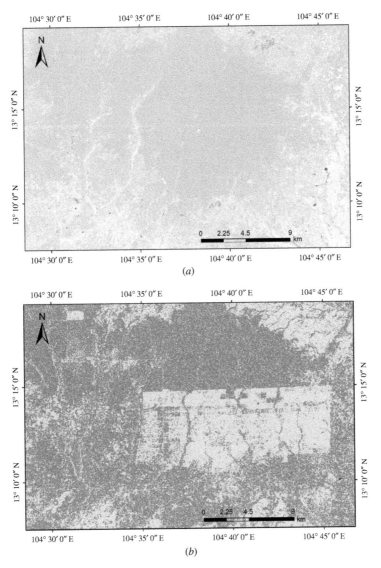

Figure 11. Change in the forest area in the north of the TSL region. (*a*) The Landsat TM image acquired in 2009. (*b*) The map of the land cover in 2012 developed in this study.

2 WF-mode SAR data. The distribution characteristics of these SAR parameters were studied to assess their usefulness in identifying artificial targets and different land-cover types. The methods used to derive the distribution characteristics of artificial targets and land-cover types from the SAR data are discussed. Based on the results of the analysis, we can draw the following conclusions: (1) The HH backscattering coefficient was found to be well suited for use in detecting artificial targets in the study area. This coefficient was the most effective of the descriptors considered in differentiating artificial targets from their background, including both water surfaces and vegetation. (2) The maximum likelihood classification method can be used to obtain land-cover information for the TSL area, using both HH and HV backscattering coefficients and the Dual-*H*, Dual-*A*, and

143

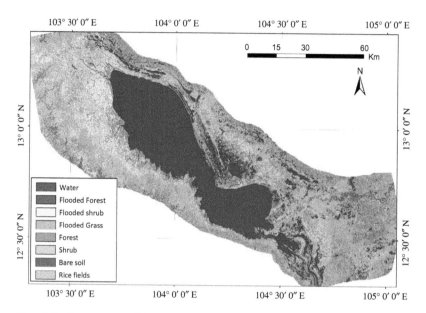

Figure 12. A land-cover map of the wetland area in TSL.

Dual-α scattering mechanism-related parameters. (3) C-band SAR data are sensitive to the small targets and are useful in target detection. However, it is difficult to differentiate flooded types from upland types because of their lower penetrating capability compared to that of long-wavelength SAR data. (4) The WF SAR data, with its high resolution and wide-swath features, can overcome the challenge of a trade-off between high data resolution and large-area coverage in image acquisition. (5) The notable expansion of human settlement and agricultural land use and the decrease of forest areas indicate the increasing human activities in the TSL region with the rapid growth of population, the development of fisheries and tourism, and lumbering.

Funding

This research was supported by the Natural Science Foundation of China [grant number 41001268]; State Key Programme of National Natural Science of China [grant number 61132006].

References

Ardila, J. P., V. Tolpekin, and W. Bijker. 2010. "Angular Backscatter Variation in L-Band ALOS ScanSAR Images of Tropical Forest Areas." *IEEE Geoscience and Remote Sensing Letters* 7 (4): 821–825. doi:10.1109/LGRS.2010.2048411.
Brisco, B., A. Schmitt, K. Murnaghan, S. Kaya, and A. Roth. 2013. "SAR Polarimetric Change Detection for Flooded Vegetation." *International Journal of Digital Earth* 6 (2): 103–114. doi:10.1080/17538947.2011.608813.
Brisco, B., R. Touzi, J. J. van der Sanden, F. Charbonneau, T. J. Pultz, and M. D'Iorio. 2008. "Water Resource Applications with RADARSAT-2–a Preview." *International Journal of Digital Earth* 1 (1): 130–147. doi:10.1080/17538940701782577.
Choi, J. H., C. W. Lee, and J. S. Won. 2009. "Flood Monitoring in Tonle Sap Floodplain using SAR Interferometry." Proceedings of 30th Asian Conference on Remote Sensing, Beijing, October.

Cloude, S. R. 2007. "The Dual Polarization Entropy/Alpha Decomposition: A PalSAR Case Study." Proceedings of PolinSAR2007, Frascati, January.

Cloude, S. R., and E. Pottier. 1996. "A Review Of Target Decomposition Theorems in Radar Polarimetry." *IEEE Transactions on Geoscience and Remote Sensing* 34 (2): 498–518. doi:10.1109/36.485127.

Cloude, S. R., and E. Pottier. 1997. "An Entropy Based Classification Scheme for Land Applications of Polarimetric SAR." *IEEE Transactions on Geoscience and Remote Sensing* 35 (1): 68–78. doi:10.1109/36.551935.

Ferro-Famil, L., A. Reigber, and E. Pottier. 2005. "Nonstationary Natural Media Analysis from Polarimetric SAR Data Using a Two-Dimensional Time-Frequency Decomposition Approach." *Canadian Journal of Remote Sensing* 31 (1): 21–29. doi:10.5589/m04-062.

Ferro-Famil, L., A. Reigber, E. Pottier, and W. M. Boerner. 2003. "Scene Characterization Using Subaperture Polarimetric SAR Data." *IEEE Transactions on Geoscience and Remote Sensing* 41 (10): 2264–2276. doi:10.1109/TGRS.2003.817188.

Frappart, F., K. D. Minh, J. L'Hermitte, A. Cazenave, G. Ramillien, T. LeToan, and N. Mognard-Campbell. 2006. "Water Volume Change in the Lower Mekong from Satellite Altimetry and Imagery Data." *Geophysical Journal International* 167: 570–584. doi:10.1111/j.1365-246X.2006.03184.x.

Gabele, M., and G. Krieger. 2008. "GMTI Performance of a High Resolution Wide Swath SAR Operation Mode." Proceedings of 2008 International Geoscience and Remote Sensing Symposium, Boston, MA, July. doi:10.1109/IGARSS.2008.4779338.

Guo, H. 2000. *Radar Earth Observation Theory and Applications*. Beijing: Science Press.

Guo, H., and X. Li. 2011. "Technical Characteristics and Potential Application of the New Generation SAR for Earth Observation." *Chinese Science Bulletin (Chinese Version)* 56: 1155–1168. doi:10.1360/972010-2458.

Kuenzer, C. 2013. "Threatening Tonle Sap: Challenges for Southeast Asia's Largest Freshwater Lake." Pacific Geographies #40, Department of Human Geography of Hamburg University, Hamburg, July/August, 29–31. ISSN 2196-1468.

Kuenzer, C., I. Campbell, M. Roch, P. Leinenkugel, V.Q. Tuan, and S. Dech. 2013. "Understanding the Impact of Hydropower Developments in the Context of Upstream–Downstream Relations in the Mekong River Basin." *Sustainability Science* 8 (4): 565–584. doi:10.1007/s11625-012-0195-z.

Kuenzer, C., H. Guo, J. Huth, P. Leinenkugel, X. Li, and S. Dech. 2013. "Flood Mapping and Flood Dynamics of the Mekong Delta: ENVISAT-ASAR-WSM Based Time Series Analyses." *Remote Sensing* 5: 687–715. doi:10.3390/rs5020687.

Kuenzer, C., H. Guo, I. Schlegel, V. Q. Tuan, X. Li, and S. Dech. 2013. "Varying Scale and Capability of Envisat ASAR-WSM, TerraSAR-x Scansar and TerraSAR-x Stripmap Data to Assess Urban Flood Situations: A Case Study of the Mekong Delta in Can Tho Province." *Remote Sensing* 5: 5122–5142. doi:10.3390/rs5105122.

Lee, J. S., and E. Pottier. 2009. *Polarimetric Radar Imaging: From Basics to Applications*. London: Taylor & Francis.

Leinenkugel, P., C. Kuenzer, and S. Dech. 2013. "Comparison and Enhancement of MODIS Cloud Mask Products for Southeast Asia." *International Journal of Remote Sensing* 34 (8): 2730–2748. doi:10.1080/01431161.2012.750037.

Leinenkugel, P., C. Kuenzer, N. Oppelt, and S. Dech. 2013. "Characterisation of Land Surface Phenology and Land Cover Based on Moderate Resolution Satellite Data in Cloud Prone Areas – A Novel Product for the Mekong Basin." *Remote Sensing of Environment* 136: 180–198. doi:10.1016/j.rse.2013.05.004.

Marko, K., M. Kummu, A. Salmivaara, S. Paradis, H. Lauri, H. Moel, P. Ward, and P. Sokhem. 2011. "Baseline Results from Hydrological and Livelihood Analyses, Exploring Tonle Sap Futures Study." Accessed December 2011. http://www.mpowernetwork.org/Knowledge_Bank/Key_Reports/PDF/Research_Reports/Exploring_Tonle_Sap_ Futures_Baseline_Report.pdf

Mather, P., and B. Tso. 2010. *Classification Methods for Remotely Sensed Data*. Boca Raton, FL: CRC Press.

Migliaccio, M., F. Nunziata, A. Montuori, X. F. Li, and W. G. Pichel. 2011. "A Multifrequency Polarimetric SAR Processing Chain to Observe Oil Fields in the Gulf of Mexico." *IEEE Transactions on Geoscience and Remote Sensing* 49 (12): 4729–4737. doi:10.1109/TGRS.2011.2158828.

Milne, T., and I. J. Tapley. 2004. "Assessment of Wetland Ecosystems and Flooding in the Tonle Sap Basin, Cambodia, Using AIRSAR." Proceedings of 2004 International Geoscience and Remote Sensing Symposium. Anchorage, September, vol. 3, 1858–1861. doi:10.1109/IGARSS.2004.1370701.

Naeimi, V., P. Leinenkugel, D. Sabel, W. Wagner, H. Apel, and C. Kuenzer. 2013. "Evaluation of Soil Moisture Retrieval from the ERS and Metop Scatterometers in the Lower Mekong Basin." *Remote Sensing* 5: 1603–1623. doi:10.3390/rs5041603.

Penny, D. 2006. "The Holocene History and Development of the Tonle Sap, Cambodia." *Quaternary Science Reviews* 25: 310–322. doi:10.1016/j.quascirev.2005.03.012.

Sarkkula, J., and J. Koponen. 2003. *Modelling Tonle Sap for Environmental Impact Assessment and Management Support*. Final Report of MRCS/WUPFIN Project. Phnom Penh: Mekong River Commission.

Sithirith, M. 2006. *The Environmental Management in Tonle Sap Lake, Cambodia*, 1–25. Phnom Penh: Fisheries Action Coalition Team (FACT).

Van Trung, N., J. H. Choi, and J. S. Won. 2009. "Landcover Change Detection at Tonle Sap, Cambodia, Using Alos Palsar." Proceedings of 30th Asian Conference on Remote Sensing, Beijing, October.

Van Trung, N., J. H. Choi, and J. S. Won. 2010. "Fusion of ALOS PALSAR and ASTER Data for Land-Cover Classification at Tonle Sap Floodplain, Cambodia." *Proceedings of SPIE* 7858: 785815-1.

Van Trung, N., J. H. Choi, and J. S. Won. 2011. "Monitoring Floodplain Area of Tonle Sap Lake, Cambodia Using Multi-Temporal ALOS PALSAR Data." *Proceedings of 2011 3rd International Asia-Pacific Conference on Synthetic Aperture Radar (APSAR)*, September, Seoul, 1–7. Piscataway, NJ: IEEE Press.

Zhang, L., H. D. Guo, and C. M. Han. 2007. "SAR Ocean Stationary Targets Detection." *Remote Sensing Technology and Application* 22 (3): 321–325.

The Ha Tien Plain – wetland monitoring using remote-sensing techniques

Tim Funkenberg[a], Tran Thai Binh[b], Florian Moder[c], and Stefan Dech[d]

[a]Institute for Geography, Julius-Maximilians-University Würzburg, Würzburg, Germany; [b]Ho Chi Minh City Institute of Resources Geography, under the Auspices of the Vietnamese Academy of Sciences, VAST-GIRS, Vietnam; [c]Ministry of Science and Technology Southern Representative Office (MOST-SRO), Vietnam; [d]German Aerospace Center (DLR), German Remote Sensing Data Center (DFD), Oberpfaffenhofen, Germany

The study investigates land-cover change in the Ha Tien Plain, once considered to be the last remaining extensive wetland area of seasonally inundated grassland in the Mekong Delta. For this purpose, two Landsat images recorded in 1991 and 2009, respectively, were classified using the C5.0 decision tree classifier. A subsequent change detection analysis revealed a significant decrease of 77% of the area classified as seasonally inundated grassland in 1991, mainly due to the conversion into agriculture, aquaculture, and forest. Since these wetlands support a high diversity of flora and a rich avifauna, there should be a focus on the protection of the last remnant patches of seasonally inundated grasslands in the Ha Tien Plain in order to preserve their biodiversity values and ecosystem functions.

1. Introduction

Wetlands belong to the most productive ecosystems in the world and deliver a wide range of benefits and values, such as flood control, shoreline stabilization, groundwater replenishment, sediment and nutrient retention and export, water purification, biodiversity protection, climate regulation, and food supply (Millennium Ecosystem Assessment 2005). The Ramsar Convention distinguishes 42 wetland types that are grouped into marine/coastal wetlands, inland wetlands, and human-made wetlands. These wetlands occur wherever water is the primary factor controlling the environment and its associated fauna and flora irrespective of whether the water is permanent or temporary, static or flowing, fresh, brackish, or salty (Ramsar Convention Secretariat 2013). The Mekong Delta, located in Southern Vietnam, spans an area of approximately 40,000 km^2 and is characterized by a steady interplay between floods from the Mekong River, heavy rains during the wet season, and tidal salt water intrusion and distinctive droughts during the dry season (Vo 2012). Owing to this hydrological influence, the Mekong Delta supports a vast range of wetland types from all categories (Vietnam Environmental Protection Agency 2005). However, the expansion and intensification of agricultural production, particularly for rice cultivation and aquaculture, mainly driven by economic reforms initiated in 1986, known as *doi moi* (renovation), have significantly reduced the proportion of natural wetlands in the delta (Torell and Salamanca 2003). Former environmental policies that have considered natural wetlands as wasteland have additionally favoured this process (Buckton et al. 1999; Vietnam Environmental Agency 2005).

Monitoring the distribution of wetlands and their dynamics is of utmost importance to support decision-making in the field of land management and nature conservation. Since wetlands are often difficult to access and field surveys are time and cost intensive, remote-sensing techniques that use satellite images, covering large geographic areas at a constant interval, provide optimal means to investigate land-cover change over time in a spatially comprehensive manner. An overview of remote sensing in conjunction with wetlands is provided in Ozesmi and Bauer (2002), Silva et al. (2008), and Adam, Mutanga, and Rugege (2010).

This study focuses on the analysis of land-cover change in the Ha Tien Plain, with an emphasis on seasonally inundated grassland. Already in 1999, Buckton et al. pointed out that the seasonally inundated grassland in the Ha Tien Plain should have the highest priority for biodiversity conservation because it represents the last remaining extensive wetland of this kind in the Mekong Delta and meets several criteria to be designated as a Ramsar site. Since the mid-1990s this unique wetland type has been increasingly threatened by the conversion into agriculture, forest plantations, and aquaculture (Buckton et al. 1999; Triet et al. 2000; BirdLife International 2004). To investigate this conversion, two Landsat 5 satellite images recorded in 1991 and 2009 have been converted to land-cover maps by a supervised classification approach using the decision tree classifier C5.0 (Quinlan 1993). A subsequent post-classification change detection approach was applied to identify the spatial distribution of land-cover change to quantify the loss of seasonally inundated grassland in the study area.

2. Study area

The study area is located in the north-western part of the Mekong Delta and lies within the administrative area of Kien Giang province (Figure 1). The study area shares its border with the city of Ha Tien and the three districts Hon Dat, Giang Than, and Kien Luong, excluding the islands in the Gulf of Thailand. The area covers 1996 km^2 and almost entirely encompasses the Ha Tien Plain, which belongs to the Long Xuyen Quadrangle, a region gently sloping towards the coast so that floodwater from the Mekong River that covers the floodplain to a depth of up to 2 m can easily drain into Gulf of Thailand (Buckton et al. 1999). The area is relatively flat and seldom exceeds an altitude of 4 m above sea level, with the exception of the precipitous limestone hills near the coastline that reach a level of up to 200 m altitude (Hon Chong). Numerous canals forming a dense network and the only major river, the Giang Thanh, runs along the Cambodian border and reaches the Gulf of Thailand via the Dong Ho tidal Lagoon. Acid-sulphate soils, which can be a source of toxic aluminium or hinder the phosphorus uptake of plants when oxidized, are predominant (Buckton et al. 1999). Rainfall varies between 1500 and 2500 mm per year, with an increasing gradient from west to east, and occurs mainly during the wet season from May to November (Le Cong 1994). Until 1994, natural or semi-natural wetland ecosystems, which were only marginally affected by anthropogenic influences, covered most of the area (Triet et al. 2000). The primary vegetation types, according to field surveys conducted during 1997 (Triet et al. 2000), were (1) seasonally inundated grassland, building diverse plant communities including 94 grass and sedge species, (2) *Melaleuca* forest, mostly in shrubby form consisting of tress 3–6 m tall and locally reaching 10–12 m, (3) *Nypa fruticans* swamp along the Dong Ho Lagoon and Giang Thanh River with *Nypa* palms reaching up to 8 m, (4) mangrove stands along the shoreline, and (5) specific karst vegetation on limestone hills. Apart from the limestone

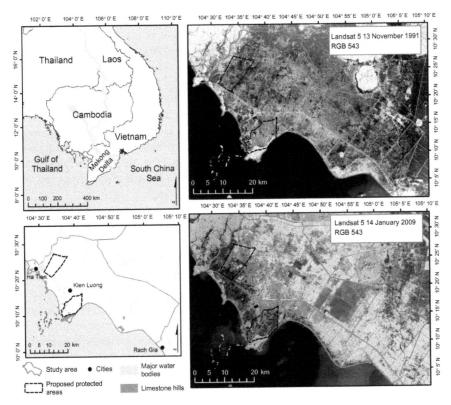

Figure 1.   Study area and satellite images used for change detection analysis.

hills with its karst formations (Deharveng et al. 2004; Truong et al. 2004), the seasonally inundated grassland is especially of high conservation value because it supports a high diversity in flora and a rich avifauna including several threatened bird species, such as Sarus Crane (*Grus antigone*), White-shouldered Ibis (*Pseudibis davisoni*), Spot-billed Pelican (*Pelecanus philippensis*), Painted Stork (*Mycteria leucocephala*), and Bengal Florican (*Houbaropsis bengalensis*) (Buckton et al. 1999; Triet et al. 2000).

3.   Data

Since the study area is situated in a subtropical, maritime region, the availability of appropriate satellite images is substantially restricted by frequent and persistent cloud cover. In addition, the number of usable satellite images is further diminished by seasonal inundation covering extensive areas of the study area during the flood season from July to December. The study is based on satellite data from the Landsat programme since it provides one of the longest archives in remote sensing and the data are free of charge, cover the study area by a single scene, and have a spatial and spectral resolution sufficient for the intended purpose. After an extensive inspection of the Landsat archive (http:// glovis.usgs.gov/), two satellite images acquired by the Thematic Mapper (TM) sensor on the Landsat 5 satellite were regarded as being suitable for a land-cover change analysis. The two selected scenes were recorded on 14 January 2009 and 13 November 1991, respectively, and are processed to level 1T, and provide six bands in the visible, near-infrared, and shortwave infrared (VNIR/SWIR) sections of the electromagnetic spectrum at 30 m spatial resolution and an additional thermal band at 120 m resolution. The 2009 scene is cloud free whereas the latter shows some clouds in the western and southeastern part of the study area. Unfortunately the image from 1991 is also partly influenced by floodwaters pushing from the north into the study area, which has to be taken into consideration when analysing land-cover change.

Furthermore, a digital elevation model (DEM) derived from the NASA Shuttle Radar Topographic Mission (SRTM) with a resolution of 3 arc seconds (approx. 90 m at the equator) was acquired from the Consultative Group on International Agricultural Research – Consortium for Spatial Information (CGIAR-CSI), with the aim to enable post-classification techniques based on altitude and inclination. The freely available DEM was preprocessed by the CGIAR-CSI with a hole-filling algorithm described by Reuter, Nelson, and Jarvis (2007) to provide a seamless and complete product.

4. Methods

4.1. Image preprocessing

The two Landsat scenes have already been processed to level 1T by the United States Geological Survey (USGS), providing systematic radiometric and geometric accuracy by incorporating ground control points while employing a digital elevation model (DEM) for topographic acuracy. A thorough comparison of the positional accuracy between the two scenes based on linear features such as canals reveals an adequate spatial match so that no further geometric correction was necessary.

To transfer digital numbers into surface reflectance values that enable the differentiation of specific land-cover types by spectral signatures (e.g. vegetation, water, bare ground), the two images were atmospherically corrected using the ATCOR 2 software (Richter 1996). Atmospheric correction is applied to rectify distortions of the electromagnetic radiation captured at the satellite caused by absorption and scattering processes in the atmosphere. Although atmospheric correction is not absolutely mandatory for a post-classification comparison change detection analysis, results are expected to be more accurate on corrected images (Song et al. 2001). Subsequently eight indices were derived from the six atmospherically corrected VNIR/SWIR bands (Table 1). The addition of indices to the spectral bands proved to enhance the discrimination of land-cover types (Davranche, Lefebvre, and Poulin 2010) and includes amongst others the widely used normalized difference vegetation index (NDVI) and the normalized difference water index (NDWI).

The DEM has been re-sampled to 30 m to match the resolution of the Landsat images. Furthermore the inclination of each pixel has been calculated from the re-sampled DEM.

4.2. Image classification

The two Landsat images were separately classified by a supervised, pixel-based classification approach, based on the C5.0 decision tree classifier (Quinlan 1993), embedded in the twinned object and pixel-based automated classification chain (TWOPAC) (Huth et al. 2012). Using C5.0 within the processing framework of TWOPAC has the advantage of significantly reducing the manual classification workload, since several steps, such as the random separation of the sample data set into training and validation data, the construction of the decision tree, as well as the validation of the classification output, are automatically

Table 1. Bands used as input for image classification. Normalized difference vegetation index (NDVI), normalized difference water index (NDWI).

No.	Formula
B1	Landsat 1
B2	Landsat 2
B3	Landsat 3
B4	Landsat 4
B5	Landsat 5
B6	Landsat 7
B7	(B4–B3)/(B4 + B3) [NDVI]
B8	B4/B6
B9	B4/B5
B10	(B4–B5)/(B4 + B5) [NDWI]
B11	B5/(B4/B6)
B12	B2/B3
B13	B5/B3
B14	B4/B2

generated. Furthermore, TWOPAC is predominantly based on open source software and offers, as a plug-in for Quantum GIS, a user-friendly graphical interface (Huth et al. 2012).

Decision trees for remote-sensing classification have already been proved to generate reasonable results (Brown de Colstoun et al. 2003; Pal and Mather 2003; Im and Jensen 2005; Wright and Gallant 2007; Kandrika and Roy 2008; Davranche, Lefebvre, and Poulin 2010; Otukei and Blaschke 2010) and have been preferred over other classification algorithms because they are relatively robust, insensitive to noisy input data, and make no assumptions about the frequency distribution of training samples (Hansen, Dubayah, and DeFries 1996; Friedl and Brodley 1997). In general a decision tree consists of several nodes or branches that recursively split the pixels of an image into smaller subdivisions until a terminal leaf, which represents the class label, is reached. The split at each node is based on a rule derived from the training data using a statistical procedure (Friedl and Brodley 1997). In the case of the C5.0 classifier, the so-called information gain ratio, which measures the reduction in entropy in the data at each split, is implemented for this purpose. At each node the attribute is chosen that maximizes the information gain ratio. The splitting proceeds until each leaf contains only training data from a single class or no gain in information is yielded by further splits (Friedl, Brodley, and Strahler 1999). This procedure often results in a very large and complex tree prone to over-fit the training data. Over-fitting is counteracted by error-based pruning described in detail by Quinlan (1993). To enhance the classification accuracy, adaptive boosting was applied, which iteratively creates a series of decision trees from the training data. Thereby each successive tree focuses on the cases that have been misclassified by the previous one. At each iteration, the training data are weighted according to their classification error. In this way, special emphasis is placed on the cases being more difficult to classify (Friedl, Brodley, and Strahler 1999). For each Landsat image, 10 iterations, with a subsequent majority vote to determine the class label of each pixel, were applied. According to Quinlan (1996), the improvement of the classification accuracy tends to stabilize after 10 iterations.

The sample data set to train the classifier and to validate the classification results is based on high-resolution satellite images from Google Earth and visual impressions from a field trip conducted in February 2012. Furthermore, specific information about land-cover types in the study area was provided by Triet et al. (2000). However, detailed ground truth data, e.g. collected during an extensive field survey, were not available. The classification scheme was initially based on spectral classes, which were subsequently merged to more generalized information classes representing (1) forest, (2) seasonally inundated grassland, (3) artificial surface, (4) agriculture, and (5) water (Tables 2 and 3). More generalized classes are used because of the lack of detailed ground truth data and because it has the intention to reduce land-cover changes caused by misclassifications. The sample data set for each spectral class has been thoroughly collected from the satellite images on a pixel-based approach taking the contextual and spectral information inherent to the image as well as the external sources into account. In this manner 80–280 pixels for each spectral class have been selected, of which 2/3 were randomly chosen for training and 1/3 for validation purposes.

4.3. Post-classification methods

Spectral confusion occurred between the 'agriculture I' spectral class and the 'forest III' spectral class, representing karst vegetation on limestone hills. To reduce these errors, all

Table 2. Classification scheme for the 1991 image.

Code	Spectral classes	Information classes	Description
1	Forest I	Forest	Forest and forest-like land-cover types (e.g.
2	Forest II		*Melaleuca, Nypa* palm, mangroves, fruit
3	Forest III		trees, and vegetation on limestone hills).
4	Forest IV		
5	Grassland I	Seasonally inundated	Grass communities according to salinity and
6	Grassland II	grassland	soil type.
7	Artificial surface I	Artificial surface	Urban areas, construction sites, quarries,
8	Artificial surface II		dikes, and bare areas.
9	Artificial surface III		
10	Agriculture I	Agriculture	Paddy fields and crops, including ploughed
11	Agriculture II		fields.
12	Water I	Water	Coastal and inland waterbodies, canals, rivers,
13	Water II		and flooded areas.
14	Water III		
15	Water IV		

Table 3. Classification scheme for the 2009 image.

Code	Spectral classes	Information classes	Description
1	Forest I	Forest	Forest and forest-like land-cover types (e.g.
2	Forest II		*Melaleuca, Nypa* palm, mangroves, fruit
3	Forest III		trees, and vegetation on limestone hills).
4	Forest IV		
5	Grassland I	Seasonally inundated	Grass communities according to salinity and
6	Grassland II	grassland	soil type.
7	Artificial surface I	Artificial surface	Urban areas, construction sites, quarries,
8	Artificial surface II		dikes, and bare areas.
9	Agriculture I	Agriculture	Paddy fields and crops, including ploughed
10	Agriculture II		fields.
11	Water I	Water	Coastal and inland waterbodies, canals, and
12	Water II		rivers.

pixels classified as 'agriculture I' and either being located at an altitude above 25 m or exceeding a maximum inclination of 12° were relabelled as 'forest III'. On the other hand, all pixels classified as 'forest III' and being situated below an altitude of 7 m or below an inclination of 12° were relabelled as 'agriculture I'. This method seems appropriate, since the karst vegetation is confined to the steep limestone hills, being the only significant topographic elevation in the study area. Furthermore only 0.48% of the total pixels in 1991 and 1.35% in 2009 are affected.

After merging the spectral classes to the five information classes, a filter was applied to smooth the classification by relabeling isolated pixels. This was carried out by associating all patches smaller than or equal to an area of 0.09 ha (corresponding to a single pixel) to the adjacent patch with which it shares the longest border. Pixels belonging to the water class or the artificial surface class were excluded from filtering to preserve linear features such as canals and roads.

4.4. *Change detection analysis*

The post-classification comparison change detection approach chosen in this study implies that the two Landsat images are classified separately from each other to subsequently detect changes of land cover by a pixel-by-pixel comparison (Coppin et al. 2004). With this technique a matrix of change directions can be produced, providing information about the gain, loss, and stability of each land-cover type in a quantitative and qualitative way (Lu et al. 2004). Coupled with a geographic information system (GIS), this information can be used to draw a comprehensive picture of the spatiotemporal land-cover change in a specific area.

Since no assumptions on the current land-cover distribution for inundated areas can be made, flooded areas in the 1991 scene were excluded from the change analyses. Temporally consistent waterbodies, such as canals, the Dong Ho Lagoon or the Giang Thanh River, however, have been retained. Water pixels in the 2009 scene were not affected by floods at all and have been entirely included in the analysis. This assumption is confirmed by the acquisition date, which is out of the flood season and visual comparisons with other Landsat images that mainly indicate constant waterbodies in the 2009 scene.

4.5. *Accuracy assessment*

To validate the classification results, the widely used overall accuracy, kappa coefficient, producer's accuracy (PA), and user's accuracy (UA) have been calculated from an error matrix for each land-cover map based on the pixels from the sample data set retained for validation purposes (Liu, Frazier, and Kumar 2007; Congalton and Green 2009).

5. Results

5.1. *Land-cover maps and change detection results*

The two land-cover maps derived from the Landsat images are illustrated in Figure 2, whereas the respective land-cover proportion of each land-cover type is shown in Figure 3. Visual inspection reveals that the study area was mainly covered by grassland, forest, and water in 1991. However, a substantial part of water consists of temporarily flooded areas, especially in the northern and western part of the study region. Agriculture

153

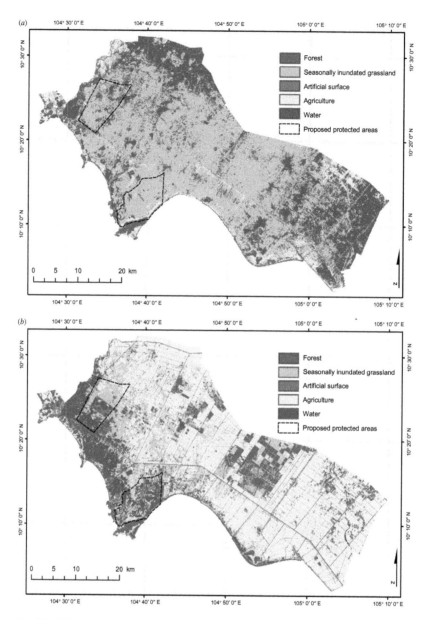

Figure 2. Classification results. (a) Land-cover classification 1991 and (b) land-cover classification 2009. Proposed protected areas are downloaded from www.protectedplanet.net and are in accordance with the areas recommended by Buckton et al. (1999).

in 1991 was confined to the northeastern, southeastern, northwestern, and western fringes of the study area, along the canal connecting Ha Tien with Rach Gia and around the limestone hills. In 2009, a strong increase in agriculture, mainly at the expense of seasonally inundated grassland and flooded areas, was evident. An expansion of waterbodies in the western and southwestern part was also apparent. As mentioned before these waterbodies are unaffected by floods and can predominantly be attributed to aquacultural ponds for fish and shrimp production (own observations, comparison with Google Earth). Furthermore, a large forest

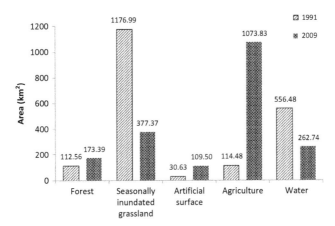

Figure 3. Land-cover proportion of each land-cover type in 1991 and 2009 as seen in Figure 2. In 1991 an additional area of 5.7 km² was covered by clouds.

plantation was established in the central part of the area. The canal system was also extended, indicated by a dense network of linear features classified as artificial surface representing roads, dikes, and buildings along the canals.

The findings from the visual inspection are in accordance with the figures of the land-cover matrix (Tables 4 and 5). Over 77% (913.04 km²) of the area classified as seasonally inundated grassland in 1991 has been converted to other land-cover types (cf. Figure 4). Over 64% (589.79 km²) of this loss is attributed to agriculture followed by nearly 17% (154.15 km²) owing to water (mainly aquaculture). These two classes also represent the land-cover types that exhibit the most relative and absolute net increase of over 506% (580 km²) in the case of agriculture and over 781% (189 km²) in the case of water. This expansion also reduced the forest area in the region, which lost over 30% of its former area to agriculture and almost 21% to water (mainly aquaculture). Owing to the conversion of 110.80 km² of grassland to forest, this class still exhibits a net increase of 42.07 km². Furthermore, artificial surfaces show a net increase of 55.97 km², predominantly on the expanse of seasonally inundated grassland. These are the main trends of land-cover change

Table 4. Land-cover matrix showing the change of each land-cover type between 1991 and 2009 (in %). Areas flooded in 1991 are excluded. Bold values indicate stable areas.

		1991				
		Forest	Seasonally inundated grassland	Artificial surface	Agriculture	Water	Class total
2009	Forest	**32.45**	9.41	1.26	3.16	13.59	10.60
	Seasonally inundated grassland	17.30	**22.43**	12.59	18.35	2.22	21.17
	Artificial surface	3.84	4.95	**38.84**	9.24	6.24	5.94
	Agriculture	26.89	50.11	29.68	**57.30**	0.61	47.63
	Water	19.53	13.10	17.64	11.95	**77.34**	14.67
	Class total	7.71	80.67	2.10	7.85	1.66	100.00
	Class changes	67.55	77.57	61.16	42.70	22.66	
	Image difference	37.38	−73.76	182.77	506.98	781.18	

Table 5. Land-cover matrix showing the change of each land-cover type between 1991 and 2009 (in km²). Areas flooded in 1991 are excluded. Bold values indicate stable areas.

		1991				
	Forest	Seasonally inundated grassland	Artificial surface	Agriculture	Water	Class total
2009 Forest	**36.52**	110.80	0.38	3.62	3.30	154.62
Seasonally inundated grassland	19.47	**263.95**	3.85	21.01	0.54	308.82
Artificial surface	4.32	58.30	**11.89**	10.58	1.52	86.60
Agriculture	30.26	589.79	9.09	**65.60**	0.15	694.89
Water	21.98	154.15	5.40	13.68	**18.78**	214.00
Class total	112.55	1176.99	30.63	114.48	24.29	1458.94
Class changes	76.03	913.04	18.73	48.89	5.50	
Image difference	42.07	−868.17	55.97	580.41	189.71	

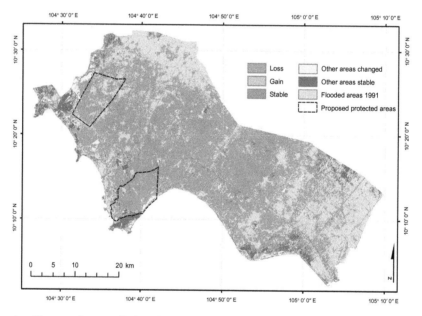

Figure 4. Change of seasonally inundated grassland from 1991 to 2009.

revealed by the land-cover matrix. However, seasonally inundated grassland despite the significant loss of its former area also shows some smaller gains (21.01 km² from the conversion of agriculture and 19.47 km² from the conversion of forest).

5.2. *Accuracy assessment*

Results of the accuracy assessment, calculated for the boosted classification including all spectral classes, are shown in Tables 6 and 7. The overall accuracy reaches 95.55% and 94.32% for the 1991 and 2009 classification, respectively. User's accuracy for the single classes varied from 85.71% to 100% in 1991 and from 88.89% to 97.73% in 2009.

Table 6. Confusion matrix 1991 with overall accuracy, producer's accuracy (PA), and user's accuracy (UA) (in %) for the boosted classification. Numbers indicate spectral classes described in Table 2. Bold values indicate correctly classified validation pixel.

		Validation pixel																
		1	2	3	4	5	6	7	8	9	10	11	12	13	14	15	Total	UA
Classified pixel	1	**48**	8	0	0	0	0	0	0	0	0	0	0	0	0	0	56	85.71
	2	4	**46**	0	2	0	0	0	0	0	1	0	0	0	0	0	53	86.79
	3	0	0	**34**	0	0	0	0	0	0	0	1	0	0	0	0	35	97.14
	4	4	0	0	**25**	0	0	0	0	0	0	0	0	0	0	0	29	86.21
	5	0	0	0	0	**65**	0	0	0	0	0	0	0	0	0	0	65	100
	6	1	0	0	0	0	**83**	0	0	0	0	0	0	0	0	0	84	98.81
	7	0	0	0	0	0	0	**26**	0	0	0	0	0	0	0	0	26	100
	8	0	0	0	0	0	0	1	**27**	0	0	0	0	0	0	0	28	96.43
	9	0	0	0	0	0	0	0	0	**27**	0	0	0	0	0	0	27	100
	10	0	0	0	0	0	2	0	0	0	**35**	1	0	0	0	0	38	92.11
	11	1	0	0	0	0	0	0	0	0	1	**42**	0	0	0	0	44	95.45
	12	0	0	0	0	0	0	0	0	0	0	0	**27**	0	0	0	27	100
	13	0	0	0	0	0	0	0	0	0	0	0	0	**27**	0	0	27	100
	14	0	0	0	0	0	0	0	0	0	0	0	0	0	**41**	0	41	100
	15	0	0	0	0	0	0	0	0	0	0	0	0	0	0	**27**	27	100
	Total	58	54	34	27	65	85	27	27	27	37	44	27	27	41	27	607	
	PA	82.76	85.19	100	92.59	100	97.65	96.30	100	100	94.59	95.45	100	100	100	100		

Overall accuracy 95.55%
Kappa coefficient 0.95

Table 7. Confusion matrix 2009 with overall accuracy, producer's accuracy (PA), and user's accuracy (UA) (in %) for the boosted classification. Numbers indicate spectral classes described in Table 3. Bold values indicate correctly classified validation pixel.

Classified pixel	Validation pixel												Total	UA
	1	2	3	4	5	6	7	8	9	10	11	12		
1	**58**	1	0	3	0	0	0	0	0	1	0	0	63	92.06
2	2	**41**	0	0	0	0	0	0	0	0	0	0	43	95.35
3	0	0	**34**	0	0	0	0	0	0	2	0	0	36	94.44
4	1	1	0	**24**	0	1	0	0	0	0	0	0	27	88.89
5	0	0	0	0	**40**	1	0	0	0	0	0	0	41	97.56
6	0	0	0	0	0	**25**	0	0	0	2	0	0	27	92.59
7	0	0	0	0	0	0	**33**	1	1	0	0	0	35	94.29
8	0	0	0	0	0	0	0	**28**	0	0	0	0	28	100
9	0	0	0	0	3	0	1	2	**43**	0	0	0	49	87.76
10	0	1	3	0	0	0	0	0	0	**87**	0	0	91	95.60
11	0	0	0	0	0	0	0	0	0	0	**33**	1	34	97.06
12	0	0	0	0	0	0	0	0	0	0	1	**36**	37	97.30
Total	61	44	37	27	43	27	34	31	44	92	34	37	511	
PA	95.08	93.18	91.89	88.89	93.02	92.59	97.06	90.32	97.73	94.57	97.06	97.3		

Overall accuracy 94.32%
Kappa coefficient 0.94

Comparison of the results obtained from an ordinary C5.0 decision tree with the ones obtained from a boosted tree reveals an overall accuracy increase of 2.8% and 1.76% in 1991 and 2009, respectively.

6. Discussion

The accuracy assessment produced fairly good results (Tables 6 and 7), which are supposed to be even better after the aggregation of the spectral classes to the five information classes. The results, however, can be expected to be slightly overestimated since the validation data were randomly selected from the sample data set within the automatic validation procedure of TWOPAC. Therefore spectral characteristics of the training and validation data exhibit high similarities *a priori*, which in turn favours high accuracies.

Besides the accuracy assessment, visual inspection provides information about the quality of the classification and helps reveal some minor deficits. On the one hand, some confusion arose from land-cover types not considered in the classification scheme because they are too small in its extent or not identifiable with the available auxiliary data in the 30 m Landsat images. This was the case for (1) vegetation along canals that were assigned to the seasonally inundated grassland class and for (2) patches of forest plantations recently cleared or in the regrowth phase, which are either confused with seasonally inundated grassland or the agriculture. On the other hand, further misclassification can be expected owing to the confusion between ploughed fields and bare areas, seasonally inundated grassland, and some agricultural crops other than rice, as well as seasonally inundated grassland and patches sparsely covered with *Melaleuca* tress. These confusions seem to be responsible for a slight overestimation of seasonally inundated grassland and a slight underestimation of forest, which was the case for the 2009 map. Furthermore, these confusions might help explain the small gain of seasonally inundated grassland mainly attributed to the conversion of agriculture and forest (Table 5).

A higher thematic differentiation of land-cover types beyond the current classification scheme would have been desirable, but was not feasible due to the following limitations: (1) the lack of detailed, local expert knowledge, (2) a ground resolution of 30 m that hardly captures the gradual transitions between land-cover types in the study area, and (3) the restriction to a single Landsat image per classification, which limits the spectral resolution to six VNIR/SWIR bands and especially hampers the differentiation of land-cover types with overlapping spectral signatures (e.g. different forest types). The first issue can be addressed by extensive field surveys and the collection of *in situ* ground truth data. However, such data have to be collected closely to the acquisition date of the satellite image, which is particularly problematic when images from the past are analysed. The second issue can be resolved with the acquisition of satellite images having a higher spatial resolution (e.g. SPOT, RapidEye, WordView) but generally comes along with less ground coverage and archives covering shorter time periods. The third issue can be diminished by implementing multi-temporal analysis using satellite images from different phenological stages (Yuan et al. 2005), by using hyperspectral sensors (Barducci et al. 2009), or by applying a multi-sensor approach. Bwangoy et al. (2010), for instance, incorporated the information from optical and radar sensors in order to map wetlands of the Congo Basin. Although all these measures would facilitate a higher thematic differentiation of land-cover types and would contribute to a general improvement of the change detection analysis, they are cost and time intensive and often only applicable with restrictions.

Despite the generalized land-cover types and the exclusion of nearly 27% of the study area from the change detection analysis because of cloud cover and flooding in the 1991 scene, the results of the study are in accordance with the findings from the literature. In several studies the conversion of seasonally inundated grassland into agriculture, forest plantations, and aquaculture is mentioned to be the most severe threat to this unique wetland type (Buckton et al. 1999; Triet et al. 2000; BirdLife International 2004).

The strong decrease of seasonally inundated grassland in the Ha Tien Plain is also a direct result of the environmental policy in Vietnam. According to the Land Law (1993), wetlands predominantly fall under the category 'unused land', especially if they do not feature any forest stands. In the Mekong Delta, these areas have been designated to be converted into agriculture, forest plantations, and aquaculture in order to increase the economic value of the region and to improve the living standard of the local people (Decision 773/TTg of the Prime Minister 1994). Indirect ecosystem services provided by such wetlands have been disregarded in this context and the policy is even more questionable since the acid-sulphate soils of the region are considered to be only suitable for marginal agriculture (Buckton et al. 1999). Furthermore, the establishment of protected areas in Vietnam has been for a long time in favour of forested land (e.g. mangroves or *Melaleuca* stands), whereas non-forested wetlands, such as the seasonally inundated grassland in the Ha Tien Plain, have been neglected. Despite Vietnam's ratification of the Ramsar Convention in 1989, which required the designation of one Ramsar site, it took another 16 years to establish a second one. To date, only five Ramsar sites exist in the whole country (Ramsar Convention – Briefing Notes 2012). Although many efforts have been already undertaken to enhance wetland conservation and management, the Vietnam Environmental Protection Agency (2005) still encountered several issues and challenges, as follows: lack of specific laws concerning wetland conservation and management; a lack of enforcement measures for implementing existing laws; a lack of agreements on coordinating mechanisms between the responsible governmental authorities; and a lack of sanctions for violations of regulations on wetland conservation.

In the case of the Ha Tien Plain, the implementation of two protected areas, covering 144.6 km^2 in total, recommended by Buckton et al. (1999) (Figures 1, 2, and 4), to conserve the biodiversity value of the area, was disregarded (BirdLife International 2004). Finally, a protected area of 28.9 km^2 was established in Phu My commune in 2004, supporting the sustainable use of *Lepironia* sedges to create an income for the local people (Triet and Cains 2007). This area is now part of the Kien Giang Biosphere Reserve, established in 2006, which also includes other parts of the Ha Tien Plain (Chu Van and Brown 2013). Although these actions will contribute to conserve the last remaining seasonally inundated grassland areas, the results of the study have shown that most of this unique wetland type is already lost.

7. Conclusion

The study has investigated land-cover change in the Ha Tien Plain with a focus on seasonally inundated grassland by comparing land-cover maps derived from Landsat satellite images recorded in 1991 and 2009, respectively. The applied remote-sensing techniques have proved useful in generating land-cover maps on a regular basis at a fairly low cost and with reasonable effort. To continue this monitoring approach satellite images from the Landsat 8 Operational Land Imager (OLI) sensor, recently put into operation, can be used. Furthermore, satellite images with a higher spatial resolution as well as field surveys are recommended to facilitate a more detailed thematic differentiation between the land-cover types in the region.

However, the study reveals that large parts of the seasonally inundated grassland in the Ha Tien Plain are already lost due to the conversion into agriculture, forest, and aquaculture. Between 1991 and 2009, 913.04 km^2 or 77% of the area classified as seasonally inundated grassland in 1991 was affected by this conversion. Unfortunately, recommendations to protect larger parts of the area were not implemented. At least, the Phu May protected area seems to be appropriate to preserve a small part of this unique wetland type, while simultaneously considering the economic needs of the local community (Triet and Cains 2007). Considering the strong decrease of seasonally inundated grassland and its high biodiversity values, coupled with the provision of ecosystem services, a stronger focus should be set on the conservation of the last remnant seasonally inundated grassland patches, taking the Phu May protected area as a model.

Funding

The research undertaken for this article was funded by the German Ministry of Education and Research, BMBF.

References

Adam, E., O. Mutanga, and D. Rugege. 2010. "Multispectral and Hyperspectral Remote Sensing for Identification and Mapping of Wetland Vegetation: A Review." *Wetlands Ecological Management* 18: 281–296.

Barducci, A., D. Guzzi, P. Marcoionni, and I. Pippi. 2009. "Aerospace Wetland Monitoring by Hyperspectral Imaging Sensors: A Case Study in the Coastal Zone of San Rossore Natural Park." *Journal of Environmental Management* 90 (7): 2278–2286. doi:10.1016/j.jenvman.2007.06.033.

BirdLife International. 2004. *Sourcebook of Existing and Proposed Protected Areas in Vietnam, Second Edition*. Cambridge: BirdLife International.

Brown de Colstoun, E. C., M. H. Story, C. Thompson, K. Commisso, T. G. Smith, and J. R. Irons. 2003. "National Park Vegetation Mapping Using Multitemporal Landsat 7 Data and a Decision Tree Classifier." *Remote Sensing of Environment* 85 (3): 316–327. doi:10.1016/S0034-4257(03)00010-5.

Buckton, S. T., N. Cu, H. Q. Quynh, and N. D. Tu. 1999. "The Conservation of Key Wetland Sites in the Mekong Delta." Vietnam Programme Conservation Report. Hanoi.

Bwangoy, J. -R. B., M. C. Hansen, D. P. Roy, G. De Grandi, and C. O. Justice. 2010. "Wetland Mapping in the Congo Basin Using Optical and Radar Remotely Sensed Data and Derived Topographical Indices." *Remote Sensing of Environment* 114: 73–86. doi:10.1016/j.rse.2009.08.004.

Chu Van, C., and S. Brown. 2013. "Using Biosphere Reserve as an Integrated Planning and Management Tool: A Case Study in Kien Giang Vietnam." In *Balance-Unbalance International Conference*, edited by S. Davis, 157–164. Noosa, Australia: Noosa Biosphere Limited & CQ University Noosa.

Congalton, R. G., and K. Green. 2009. *Assessing the Accuracy of Remotely Sensed Data: Principles and Practices*. 2nd ed. Boca Raton, FL: CRC Press.

Coppin, P., I. Jonckheere, K. Nackaerts, B. Muys, and E. Lambin. 2004. "Review Article Digital Change Detection Methods in Ecosystem Monitoring: A Review." *International Journal of Remote Sensing* 25 (9): 1565–1596. doi:10.1080/0143116031000101675.

Davranche, A., G. Lefebvre, and B. Poulin. 2010. "Wetland Monitoring Using Classification Trees and SPOT-5 Seasonal Time Series." *Remote Sensing of Environment* 114 (3): 552–562. doi:10.1016/j.rse.2009.10.009.

Deharveng, L., L. C. Kiet, L. M. Man, and A. Bedos. 2004. "Hot Issues in Karst Conservation: The Biodiversity of Hon Chong Hills (Southern Vietnam), with Emphasis on Invertebrate

Endemism." In Proceedings of the International Transdisciplinary Conference on Development and Conservation of Karst Regions, edited by O. Batelaan, M. Dusar, J. Masschelein, T. T. Van, V. T. Tam, and N. X. Khien, 40. Hanoi, Vietnam. Thanh Xuan, Hanoi: Research Institute of Geology and Mineral Resources.

Decision 773/TTg of the Prime Minister. 1994. "Decision no. 773-TTg on the 21st OD December 1994 of the Prime Minister on the Program of Tapping and Using Waste Land, Alluvial Soil on River and Sea Shores, and Water Surface in the Plains." Accessed August 2013. http://www. moj.gov.vn/vbpq/en/Lists/Vn%20bn%20php%20lut/View_Detail.aspx?ItemID=2706

Friedl, M. A., and C. E. Brodley. 1997. "Decision Tree Classification of Land Cover from Remotely Sensed Data." *Remote Sensing of Environment* 61: 399–409. doi:10.1016/S0034-4257(97)00049-7.

Friedl, M. A., C. E. Brodley, and A. H. Strahler. 1999. "Maximizing Land Cover Classification Accuracies Produced by Decision Trees at Continental to Global Scales." *IEEE Transactions on Geoscience and Remote Sensing* 37 (2): 969–977. doi:10.1109/36.752215.

Hansen, M., R. Dubayah, and R. DeFries. 1996. "Classification Trees: An Alternative to Traditional Land Cover Classifiers." *International Journal of Remote Sensing* 17 (5): 1075–1081. doi:10.1080/01431169608949069.

Huth, J., C. Kuenzer, T. Wehrmann, S. Gebhardt, V.Q. Tuan, and S. Dech. 2012. "Land Cover and Land Use Classification with TWOPAC: Towards Automated Processing for Pixel- and Object-Based Image Classification." *Remote Sensing* 4 (12): 2530–2553. doi:10.3390/rs4092530.

Im, J., and J. Jensen. 2005. "A Change Detection Model Based on Neighborhood Correlation Image Analysis and Decision Tree Classification." *Remote Sensing of Environment* 99 (3): 326–340. doi:10.1016/j.rse.2005.09.008.

Kandrika, S., and P. S. Roy. 2008. "Land Use Land Cover Classification of Orissa Using Multi-Temporal IRS-P6 Awifs Data: A Decision Tree Approach." *International Journal of Applied Earth Observation and Geoinformation* 10 (2): 186–193. doi:10.1016/j.jag.2007.10.003.

Land Law. 1993. "Land Law of Vietnam 1993." Accessed August 2013. http://faolex.fao.org/docs/ texts/vie4824.doc

Le Cong, K. 1994. "Native Freshwater Vegetation Communities in the Mekong Delta." *International Journal of Ecology and Environmental Sciences* 20: 55–71.

Liu, C., P. Frazier, and L. Kumar. 2007. "Comparative Assessment of the Measures of Thematic Classification Accuracy." *Remote Sensing of Environment* 107 (4): 606–616. doi:10.1016/j. rse.2006.10.010.

Lu, D., P. Mausel, E. Brondízio, and E. Moran. 2004. "Change Detection Techniques." *International Journal of Remote Sensing* 25 (12): 2365–2401. doi:10.1080/0143116031000139863.

Millennium Ecosystem Assessment. 2005. *Ecosystems and Human Well-Being: Wetlands and Water Synthesis.* Washington, DC: World Resource Institute.

Otukei, J. R., and T. Blaschke. 2010. "Land Cover Change Assessment Using Decision Trees, Support Vector Machines and Maximum Likelihood Classification Algorithms." *International Journal of Applied Earth Observation and Geoinformation* 12 (1): S27–S31. doi:10.1016/j. jag.2009.11.002.

Ozesmi, S. L., and M. E. Bauer. 2002. "Satellite Remote Sensing of Wetlands." *Wetlands Ecology and Management* 10: 381–402. doi:10.1023/A:1020908432489

Pal, M., and P. M. Mather. 2003. "An Assessment of the Effectiveness of Decision Tree Methods for Land Cover Classification." *Remote Sensing of Environment* 86 (4): 554–565. doi:10.1016/ S0034-4257(03)00132-9.

Quinlan, J. R. 1993. *C4.5: Programs for Machine Learning.* San Mateo, CA: Morgan Kaufman.

Quinlan, J. R. 1996. "Bagging, Boosting and C4.5." Proceedings of the 13th national conference on artificial intelligence, 725–730. Portland, Oregon: AAAI Press.

Ramsar Convention – Briefing Notes. 2012. "The Annotated Ramsar List: Viet Nam Ramsar Convention." Accessed August 2013. http://www.ramsar.org/cda/en/ramsar-pubs-notes-annotated-ramsar-15775/main/ramsar/1-30-168%5E15775_4000_0.

Ramsar Convention Secretariat. 2013. *THE RAMSAR CONVENTION MANual: A Guide to the Convention on Wetlands (Ramsar, Iran, 1971).* Gland, Switzerland: Ramsar Convention Secretariat.

Reuter, H. I., A. Nelson, and A. Jarvis. 2007. "An Evaluation of Void-Filling Interpolation Methods for SRTM Data." *International Journal of Geographical Information Science* 21 (9): 983–1008. doi:10.1080/13658810601169899.

Richter, R. 1996. "A Spatially Adaptive Fast Atmospheric Correction Algorithm." *International Journal of Remote Sensing* 17 (6): 1201–1214. doi:10.1080/01431169608949077.

Silva, T. S. F., M. P. F. Costa, J. M. Melack, and E. M. L. Novo. 2008. "Remote Sensing of Aquatic Vegetation: Theory and Applications." *Environmental Monitoring and Assessment* 140: 131–145. doi:10.1007/s10661-007-9855-3.

Song, C., C. E. Woodcock, K.C. Seto, M. Pax Lenny, and S. A. Macomber 2001. "Classification and Change Detection Using LAndsat TM Data: When and How to Correct Atmospheric Effects?" Remote Sensing of Environment 75 (2): 230–244. doi:10.1016/S0034-4257(00)00169-3.

Torell, M., and A. M. Salamanca. 2003. "Wetlands Management in Vietnam's Mekong Delta: An Overview of the Pressures and Responses." In *Wetlands Management in Vietnam: Issues and Perspectives*, edited by M. Torell, B. D. Salamanca, and A. M. Ratner, 1–16. Penang, Malaysia: World Fish Center.

Triet, T., and R. Cains. 2007. "Towards Sustainable Rural Development: Combining Biodiversity Conservation with Poverty Alleviation – A Case Study in Phu My Village, Kien Giang Province, Vietnam." In *Transition to a Resource-Circulation Society: Strategies and Initiatives in Asia*, edited by T. Morioka, 181–190. Osaka: Osaka University Press.

Truong, Q. T., D. P. Diep, M. D. Hoang, B. T. Le, and P. N. Nguyen. 2004. "Biodiversity in the Limestone Area of Ha Tien and Kien Luong, Kien Giang Province." In Proceedings of the International Transdisciplinary Conference on Development and Conservation of Karst Regions, edited by O. Batelaan, M. Dusar, J. Masschelein, T. T. Van, V. T. Tam, and N. X. Khien, 233. Hanoi, Vietnam. Thanh Xuan, Hanoi: Research Institute of Geology and Mineral Resources.

Triet, T., R. J. Safford, P. Tran Duy, N. Duong Van, and E. Maltby. 2000. "Wetland Biodiversity Overlooked and Threatened in the Mekong Delta, Vietnam: Grassland Ecosystems in the Ha Tien Plain." *Tropical Biodiversity* 7 (1): 1–24.

Vo, K. T. 2012. "Hydrology and Hydraulic Infrastructure Systems in the Mekong Delta." In *The Mekong Delta System -Interdisciplinary Analyses of a River Delta*, edited by F. G. Renaud and C. Kuenzer, 49–81. Heidelberg: Springer.

Vietnam Environment Protection Agency. 2005. "Overview of Wetlands Status in Viet Nam Following 15 Years of Ramsar Convention Implementation." Hanoi, Vietnam, 72 pp.

Wright, C., and A. Gallant. 2007. "Improved Wetland Remote Sensing in Yellowstone National Park Using Classification Trees to Combine TM Imagery and Ancillary Environmental Data." *Remote Sensing of Environment* 107 (4): 582–605. doi:10.1016/j.rse.2006.10.019.

Yuan, F., K. E. Sawaya, B. C. Loeffelholz, and M. E. Bauer. 2005. "Land Cover Classification and Change Analysis of the Twin Cities (Minnesota) Metropolitan Area by Multitemporal Landsat Remote Sensing." *Remote Sensing of Environment* 98: 317–328. doi:10.1016/j.rse.2005.08.006.

Operational multi-sensor monitoring of turbidity for the entire Mekong Delta

Thomas Heege[a], Viacheslav Kiselev[a], Magnus Wettle[a], and Nguyen Nghia Hung[b]

[a]EOMAP GmbH & Co.KG, Gilching, Germany; [b]SIWRR, Southern Institute for Water Resources Research, Ho Chi Minh, Vietnam

An operational satellite-based approach was implemented to monitor turbidity and organic absorption in the Mekong river system. Using physics-based algorithms linked together in a fully automated processing chain, more than 300 Landsat Enhanced Thematic Mapper (ETM) scenes and 1000 MODIS scenes, representing five years of data, were used to produce standardized, quantitative time series of turbidity and organic absorption across Vietnam, Thailand, Cambodia, Laos, and China. To set up this system, the specific inherent optical properties (SIOPs) of the Mekong river system were determined through three separate field campaigns, laboratory analysis, and subsequent optical closure calculations. Following this, a range of satellite data types was tested using the derived Mekong-specific inherent optical properties, including Moderate Resolution Imaging Spectroradiometer (MODIS) 500 m data, Landsat ETM, Medium Resolution Imaging Spectrometer (MERIS), Satellite Pour l'Observation de la Terre (SPOT) 5, RapidEye, Advanced Spaceborne Thermal Emission and Reflection Radiometer (ASTER), and QuickBird. The satellite-based turbidity estimates were coincident with available field data, and comparisons showed them to be in good agreement. Overall, the derived SIOPs were suitable for water-quality monitoring of the Mekong, and the MODIS, MERIS, Landsat, and RapidEye sensors were found to be the most radiometrically stable and thereby suitable for ongoing operational processing. The implemented system delivers consistent results across the different satellite sensors and over time, but is limited to where the spatial resolution of the sensor is still able to resolve the river width. The system is currently applicable for the entire Mekong river system, both for near-real-time monitoring and for analysis of historical data archive.

1. Introduction

Directly affecting seven countries, with estimated potential socio-economic and environmental impacts on more than 220 million people over a 2.32 million km^2 area (Asian Development Bank 2012), the Mekong River system has a wide range of stakeholders. In addition, the Mekong River Delta is one of the most productive food regions in the world, and the approximately 18 million people living in the delta rely on it as a primary food source (Asian Development Bank 2012). The entire Mekong is under significant pressure from various sources, notably an increasing number of upstream damns that are causing a significant reduction in the downstream nutrient-rich sedimentation (Mekong River Commission MRC 2010; Kuenzer et al. 2012).

There is therefore an ongoing need to monitor the environmental status of the entire Mekong river system. To this end, two significant environmental variables that can be routinely measured from space are turbidity and related total suspended matter (TSM)

concentrations in the river. Directly linked to water quality, these variables provide quantitative information on sediment and related nutrient transports, and allow for the monitoring of the effects of environmental and physical changes along the river system.

Traditional *in situ* methods for monitoring these two environmental variables present a number of challenges and limitations when applied to long-term, regional-scale monitoring systems. An important requirement for a monitoring system is the acquisition of standardized measurements; measurements need to be directly comparable between stations, teams, and protocols, besides being comparable across national boundaries and, importantly, through time. The latter assumes that historical data are even available. For the Mekong region as well as for many parts of the world, the lack of transnational harmonized, reliable data is a challenge for natural resource planning (Irvine et al. 2011).

Turbidity, related TSM, and organic absorption in rivers can be quantitatively measured using remote sensing, and this technology can overcome both the challenge of consistently acquiring standardized, comparable measurements through space and time, and the limitations of access to historical data. However, this requires the application of a physics-based (as opposed to empirical) method (Heege, Odermatt, and Kiselev 2007), which can take into account the SIOPs of the waterbody being measured, the specifications of the instrument being used, and the impact of recording conditions such as viewing and illumination geometries, atmospheric effects, Sunglitter, and adjacency effects. In principle, such a method is therefore sensor-, location-, and time-independent, and meets the requirements for a long-term environmental monitoring system on a regional scale. In practice, accuracies will depend on the characteristics of the different sensors such as the radiometric, spatial, and spectral resolution, as well as the physical processing system itself.

In this article we present the successful implementation and validation of a physics-based, operational remote-sensing system for monitoring turbidity and organic absorption for the entire Mekong River system.

2. State of the art

Multispectral satellite sensors are capable of measuring water constituents, using sunlight in the visible region of the spectrum, which penetrates the atmosphere and waterbody. This light is absorbed and scattered as a function of the particles and dissolved materials in the waterbody, and the reflected light spectrum detected by the satellite sensors can be used to analyse the in-water and atmospheric properties. Several suitable multispectral satellite sensors are currently available, including Moderate Resolution Imaging Spectroradiometer (MODIS), Medium Resolution Imaging Spectrometer (MERIS), Landsat, RapidEye, and Worldview-2. These sensors differ in the number of spectral bands, radiometric stability and sensitivity to light, and spatial and temporal resolution. Sensors with low spatial resolution usually have a higher revisit frequency. For example, the 250 m spatial resolution MODIS sensors can provide up to twice daily coverage, whereas the 30 m spatial resolution Landsat 7 or 8 sensors have a revisit interval of approximately 14 days.

Space-based sensors such as MODIS and MERIS have been successfully applied to operational water-quality monitoring of both oceanic and coastal waters over the past several years. Existing algorithms can be grouped into three main approaches: manual, digitizing approaches such as used by Evans et al. (2012) to generate turbidity fronts from MODIS 250 m imagery, empirical and semi-analytical approaches such as applied in Sipelgas, Raudsepp, and Kõuts (2006) or Le et al. (2009), or physics-based approaches that rely on an inversion of radiance or reflectance spectra back to water optical properties and water constituents (Schroeder et al. 2007; Doerffer and Schiller 2007; Heege et al. 2009).

To begin with, manual digitizing approaches cannot deliver the quantitative information required for the applications discussed in this article. Empirical approaches typically rely on using band ratios or principal component analysis of band intensities or surface reflectances to derive water-quality information (e.g. Brezonik, Menken, and Bauer 2005; Hellweger, Miller, and Oshodi 2007). Several publications demonstrate the capabilities of satellite sensors to monitor seasonal and long-term trends (Martinez et al. 2009; Diaz-Delgado et al. 2010; Olmanson, Brezonik, and Bauer 2011; Güttler, Niculescu, and Gohin 2013) using empirical approaches. However, these applications remain limited to localized scales, often requiring *a priori* knowledge; they cannot be applied at inter-regional, transnational, or continental scales.

In contrast, physics-based spectral inversion algorithms have been shown to be capable of delivering quantitative, standardized water-quality information across all relevant spatial and temporal scales (Malthus et al. 2012), even if it is acknowledged that information on specific bio-optical properties – and how to treat variations of these within the inversion algorithms – is an area of ongoing research. Using regional, or so-called adaptive, SIOPs often significantly increases the accuracy for retrieving sensitive water-quality parameters such as phytoplankton and chlorophyll concentrations or suspended matter. However, for estimating the backward scattering and related turbidity parameter, the sensitivity is lower. Physics-based retrieved backscattering is therefore largely a standardized measure of turbidity. In contrast to empirical methods, it is therefore valid on continental scales besides being sensor-independent.

The approaches outlined above have been presented in the context of oceanic and coastal applications. Inland waters typically offer a considerably more challenging environment for satellite-based water-quality monitoring. One of the main contributing factors to this is the so-called adjacency effect. This is caused by the scattering of photons from surrounding terrestrial targets into the signal path of the remotely sensed inland water, effectively contaminating the low signal-to-noise information of the aquatic target. A second significant challenge is the increased optical complexity typical of inland waters. Rivers often exhibit turbidity levels of one or more orders of magnitudes higher than coastal environments. An additional challenge is the requirement for relatively high spatial resolution data in order to resolve features such as smaller rivers and streams. For this, satellite sensors such as Landsat 5, 7, and 8 (30 m resolution) or RapidEye (5 m resolution) may be more suitable. However, this entails a trade-off in reduced radiometric and spectral resolution and sensitivities, when compared with the coarser spatial resolution MERIS and MODIS sensors (300–500 m), which are specially designed for the observation of coastal waters.

To date, therefore, inland water-quality remote sensing has been performed using existing algorithms for coastal applications, and applied to water bodies large enough for coarser resolution sensors (Duan et al. 2012; Odermatt, Giardino, and Heege 2010). Here, we introduce a generic, physics-based approach capable of being adapted to the specific conditions of inland waters, and furthermore able to generate standardized output from a range of multispectral satellite sensors with varying resolutions (Heege et al. 2009).

3. Methodology

3.1. Data-processing methodology

The core of the methodology is the Modular Inversion and Processing System (MIP) (Heege and Fischer 2004; Heege et al. 2003; Heege 2000). This software integrates various physics-based algorithms for both shallow and optically deep inland or coastal waters to derive water-quality parameters, sea-floor properties, and bathymetry (Kobryn et al. 2013; Ohlendorf et al. 2011; Heege et al. 2009). The remainder of this article focuses on the retrieval of water-quality parameters, also referred to as water constituent concentrations.

The MIP architecture systematically manages the independent properties of sensor parameters and specific optical properties, as well as the radiative transfer relationships (at 1 nm spectral resolution) in the main database, and furthermore manages the uptake of the individual algorithm modules. This allows for an automatic adaptation of the retrieval modules to the specific sensor properties and optical conditions of the target area. Hence, the retrieval modules deliver standardized, inter-comparable results for various satellite sensors, to the extent that the radiometric calibration of these sensors is physically correct (and in agreement with the underlying physical models of the MIP), sufficiently stable, and appropriately sensitive. However, it should be noted that each sensor has different spectral and radiometric resolutions, and these limit the number of resolvable parameters on a case-by-case basis. The type and number of parameters to be resolved are therefore defined in the configuration file of the central retrieval module. Specifications for the sensors used here are described below.

Radiative transfer calculations, connecting the top-of-atmosphere radiances with the optical properties of water media, are performed using a finite element model (FEM) (Kisselev, Roberti, and Perona 1995; Bulgarelli, Kisselev, and Roberti 1999), encompassing the full bidirectional properties of the atmosphere, the air–water interface, and the waterbody itself. These are stored in the sensor-independent, so-called main database with several sub-databases: The 'L' database contains the relation between top-of-atmosphere radiances and the subsurface reflectance as a function of various atmospheric aerosol types, aerosol concentrations, and geometric characteristics such as view and Sun polar angles and azimuths. The 'Q' database contains the bidirectional relations between downward flux and upwelling radiance, and a third database comprises the diffuse radiance at the sensor level for land-reflected radiance, which is used by the adjacency correction module. From these databases, sensor-specific, global databases are created for each satellite sensor. These data are integrated over the appropriate spectral band for the relevant sensor, while all other variables are retained. For connecting the optical properties of water media with the concentrations of water species, we furthermore assume that optical properties of the water constituents are the products of corresponding specific optical properties (normalized to unit concentration) and concentrations, and use an analytical approximation from Albert and Mobley (2003) to calculate the subsurface irradiance reflectance as a function of absorption and scattering properties of the waterbody.

The retrieval modules load the sensor databases into the core memory and access further sensor definitions and target optical properties. The atmosphere is characterized by the surface-layer aerosol type and optical thickness and season. For higher levels, standard distributions of aerosol and molecular species concentrations are used. In the present version of MIP, the model of the atmospheric stratification follows the main features adopted in the MODTRAN code (Abreu and Anderson 1996). For the water model, the scattering, backscattering, and absorption coefficients of water are expressed as a linear combination of location-dependent specific optical properties of water constituents, where the concentrations of each constituent are weighting factors of the linear combination. Normally up to three water species, namely suspended matter, phytoplankton, and dissolved organic material (yellow substance), are used. The appropriate specific optical

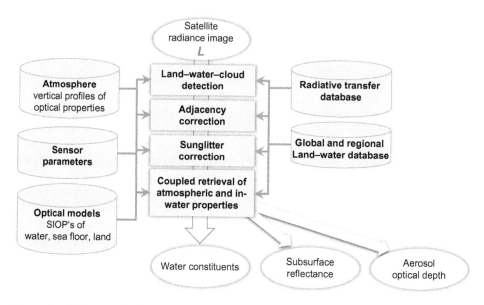

Figure 1.   Scheme of the MIP modules for water-quality processing.

properties are selected based on expert estimates, from *in situ* measurements if available, or automatically as a function of retrieved water constituent properties. The selection of the specific optical properties used here and the number of water constituents retrieved in the processor configuration for application in the Mekong are described below.

Within the water-quality processing chain (Figure 1), several modules are applied successively, covering surface-type detection (land–water–cloud detection); adjacency and Sunglitter correction; iterative retrieval of atmospheric properties; transformation to under-water reflectance; and estimation of the water constituents' scattering and absorption properties.

Surface-type detection is based on the differences of spectral trends of land, water, and cloud reflectivities. To begin with, top-of-atmosphere radiances are transformed to surface reflectivities using the L database with fixed atmospheric aerosol optical thickness (the value is 0.1). Pixels with high reflectivities in the blue region of the spectra (the threshold value depends on the channel wavelength of the sensor) are classified as cloud pixels. The discrimination between land and water pixels is made using two criteria: water pixels must have low reflectance in the near infrared region and show no increase of reflectance with wavelength in the 'red edge' region (near 700 nm). To assign mixed cloud-water pixels over water to the cloud flag, the module also considers an external global land-water database: the Shuttle Radar Topography Mission (SRTM) Water Body Data (SWBD 2005).

The adjacency correction module cycles through target water pixels. For each target pixel it calculates the sum of the contribution of the radiances reflected by neighbouring pixels and subtracts the resulting sum from the measured total radiance of the target pixel. The neighbouring pixel contributions are estimated using the analytical solution of the three-dimensional radiative transfer equation for a ground-level point source (the point spread function in the primary scattering approximation) and upward fluxes near the surface. Upward fluxes are estimated using the database storing the dependence of the top-of-atmosphere radiances on the land reflectivity, assuming 0.1 as the standard value of atmospheric aerosol optical thickness.

To increase the processing speed for satellite sensors with spatial resolutions finer than 100 m, the adjacency radiance is only calculated for every tenth pixel and then spatially interpolated for pixels in between. When using high-resolution imagery in particular, it can be necessary to account for the adjacency impact of land pixels outside the image borders. For this, a mean land albedo is calculated using the satellite scenes and a global land–water database.

The central algorithm is based on a coupled retrieval of atmospheric and aquatic properties that provide the best fit (1) between measured and modelled radiances varying atmospheric aerosol optical thickness, and (2) for each optical thickness between retrieved underwater reflectances and those modelled as a function (Albert and Mobley 2003) of the linear combination of specific optical properties and species concentrations. The number of retrieved parameters depends on the spectral and radiometric resolutions of the sensor. At least three parameters are typically retrieved, namely the suspended matter concentration (or the related backscattering coefficient), the total absorption coefficient of water constituents, and atmospheric aerosol optical thickness. With the increasing number of sensor channels, such as for example the MERIS sensor, which has 15 bands with suitable radiometric resolution, the absorbing water constituents of coloured dissolved organic material (CDOM) and phytoplankton can also be derived separately. In contrast, the two MODIS 250 m bands only allow for the retrieval of two parameters: aerosol optical depth and suspended matter concentration. In this case, the absorption properties are either initially calculated from the MODIS 500 m data (which has additional spectral channels) or, if these are not available, assigned a fixed value as defined in the configuration file.

As a standard part of the processing, an accuracy or uncertainty indicator is calculated for each retrieved parameter and for each pixel in the image data. This measure comprises a comprehensive range of factors that can impact the derived product quality, which include the geometry between Sun, target, and sensor; the estimated Sunglitter probability; the retrieved aerosol optical depth; residuals of the measured and modelled sensor radiances and subsurface reflectances; and the comparison of retrieved water species concentrations to extreme values as defined in the configuration files. The total uncertainty indicator is normalized to a value between 0 and 1 and can also be used during the generation of third-level products. When averaging or integrating parameter values through space and/or time, the uncertainty measure can be used to weight the contribution of each pixel to the new aggregate product.

To deploy MIP operationally, the EOMAP Workflow System (EWS) is used. The complete water-quality processing chain covers approximately 100 individual processing steps, from initial file import actions to final output actions that generate WGS- and INSPIRE-conform metadata for the output products. This processing chain is orchestrated and controlled by the EWS. The EWS also triggers event-based processing in fully operational environments, and is therefore readily integrated with satellite receiving stations. The EWS is an essential component for operational near-real-time remote-sensing applications and robust production of extended time series.

3.2. MIP settings and configuration

For the surface-type detection (land–water–cloud), the settings in the configuration files are related to wavelength, and bands are automatically selected according to the spectral position of the available bands of each satellite sensor. The module performs a preliminary standard atmospheric correction with an initial aerosol value of 0.1 to retrieve surface albedo values for each band. For cloud detection, the module selects the band with the shortest available wavelength (e.g. the blue channel, or band 1, of MODIS or Landsat). The threshold of the cloud detection is set to 0.6 for MODIS and to 0.35 for Landsat. The upper threshold of albedo for the first part of the water detection is set to 0.2 for MODIS band 4 at 857 nm and Landsat band 4 at 834 nm, and to 0.17 for Landsat band 5 at 1650 nm. The threshold for the minimum increase of land-surface albedo near the red-edge bands is set to 0.03 for MODIS bands 3 at 646 nm and 4 at 857 nm, and to 0.05 for Landsat 7 bands 3 and 4 at 661 and 834 nm.

The adjacency impact is calculated also for an initial aerosol concentration of 0.1 and a fixed 'maritime' aerosol type. The Sunglitter correction is applied only for MODIS data, but the algorithm is effective only over larger water areas. It is automatically deactivated for pixels in small-scale inland waters to avoid insufficient discrimination between impact of aerosols and Sunglitter. For Landsat and other high-resolution sensors, the Sunglitter correction is deactivated in the workflow system, as these sensors are spectrally not capable of discriminating between aerosol and Sunglitter. However, the Sunglitter probability is still calculated for each pixel, as a function of the geometry, and incorporated in the delivered quality measure product (described in Section 3.1).

The number and type of water constituents to be derived, the choice of specific optical properties, and any necessary adaptations to sensor-specific shortcomings (e.g. corrupt or insensitive spectral bands) are defined in the configuration files of the retrieval modules.

The configuration applied for the Mekong river independently retrieved aerosol optical depth, the scattering by suspended solids, and the sum of total dissolved and particulate organic absorption. Chlorophyll was not independently retrieved, as it does not dominate the measurable light absorption in the Mekong river and can spectrally not be further discriminated from the organic absorption with the sensors used here.

The range of aerosol optical properties was set at 0.01–0.6 m^{-1} at 550 nm for the maritime aerosol type, and at 0.0 for all other aerosols. The range of turbidity is from 0.01 to 500 NTU (nepholo-metric turbidity unit), and the yellow substance absorption from 0.01 to 155 m^{-1}. Based on fieldwork and image-processing experience, the chlorophyll absorption was fixed at a value of 5 µg l^{-1}.

Pixels with retrieved aerosol concentrations that reach a value of 0.6 over water are assigned to the cloud flag.

3.3. Specific inherent optical properties

The specific inherent optical properties (SIOPs) used for the retrieval of water-quality parameters are defined in the module configuration files. Specific water types are defined as a set of specific absorption, scattering, and backscattering coefficients for each optically active water component. The components are here defined as follows: turbidity (with a one-to-one linear relation to suspended matter for lower concentrations), dissolved organic absorbers, and chlorophyll. For chlorophyll, we used the specific absorption spectra as published in Heege (2000) and Heege and Fischer (2004), and with scattering and back-scattering values set to zero (0). Scattering of phytoplankton cells is therefore included in the scattering of suspended matter and therefore related to turbidity. The specific scattering coefficients b for turbidity and related suspended matter are plotted in Figure 2 (right-hand side), with spectra originally derived for Lake Constance in 2007. The relation between the backscattering b_b and scattering coefficients b is fixed at $b_b/b = 0.019$ (as in Heege and Fischer (2004). For the yellow substance, an absorption spectrum normalized at 440 nm to 1 with an exponential decrease, $S = 0.0085$ nm^{-1}, was derived from the data

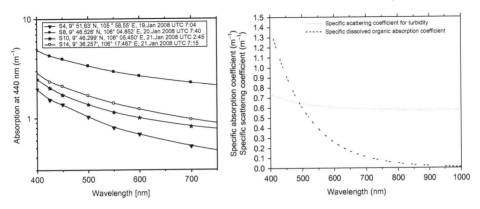

Figure 2. Examples of dissolved organic absorption spectra (left side) and specific absorption and scattering coefficients (right side).

gathered during several *in situ* campaigns in the Mekong Delta and verified using several optical closure calculations with underwater radiometer data and satellite data. For pure water absorption and scattering, spectra of Buiteveld, Hakvoort, and Donze (1994) and Smith and Baker (1981) were used.

3.4. Measurement campaigns

Remote-sensing field validation campaigns were undertaken in 2008, 2009, and 2012 in the Bassac and Can Tho rivers of the Mekong. From 19 to 21 January 2008, boat-based measurements were carried out at 15 locations in Bassac river from the city of Can Tho in the centre of the Mekong Delta, and down to the river mouth for approximately 80 km. Measurements from 15 to 20 March 2009 were undertaken in the Bassac river and tributaries at 52 stations from Can Tho to Hong Ngu, 90 km upstream and close to Cambodia. Finally, 64 measurements of Secchi depth were carried out from a boat during 24 and 25 February 2012 in the Bassac and Can Tho rivers, close to city Can Tho.

Measurements included *in situ* sampling of suspended matter, measurement of Secchi depth, dissolved organic absorption (during 2008 only), and underwater spectrometer measurements (during 2008 and 2009). The spectrometer measurements were carried out with two RAMSES irradiance spectrometers (one for downwelling and the other for upwelling light) and one RAMSES radiance spectrometer for upwelling light. Figure 3 shows the arrangement of the sensors before lowering from the booms installed at the boat. All sensor heads are located on one plane approximately 20 cm from the water surface, and used to measure the ratio between the upwelling and downwelling irradiances, the irradiance reflectance, and the remote-sensing reflectance. Suspended matter was measured gravimetrically with the dry weight of glass fibre filters by the Southern Institute of Water Related Resources SiWRR in Ho Chi Minh City. Samples to calculate the absorption of dissolved organic materials were successively filtered with 1 μm and 0.2 μm filters. Figure 2 (left-hand side graph) shows examples of the derived absorption spectra. These were used to calculate an initial spectral slope of the (mostly) dissolved organic materials, with a mean value of approximately 0.0085. The measured spectra do

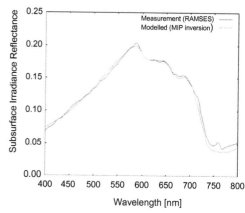

Figure 3. The RAMSES spectrometer devices above water, before lowering into the Mekong. Also shown is a derived subsurface irradiance spectrum for station 1, close to Can Tho, on 19 January 2008. The red and green lines show the RAMSES-derived spectrum and the modelled spectrum, respectively.

show a varying offset, which may be related to remaining particles in the filtered Mekong water or to artefacts of the laboratory spectrometer.

Water samples to gravimetrically measure suspended matter (SM) in the laboratory were only taken at a limited set of stations, but always in combination with the Secchi depth (SD). We used these measurements to establish a relation between SD and SM (see Figure 4), in order to calculate suspended matter from Secchi measurements for stations where no suspended matter samples were taken. The relationship established was

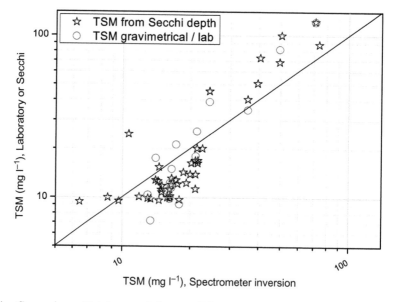

Figure 4. Comparison of total suspended matter (TSM) concentrations derived from gravimetrical analysis of *in situ* samples or Secchi disc measurements (*y*-axis) *versus* TSM concentrations calculated from the inversion of underwater reflectance spectra measurements using the established set of SIOPs (*x*-axis).

$$TSM = 9.4 + 379.3\exp(-(SD)/0.14). \tag{1}$$

Concentrations of TSM in the dry season ranged from 10 to 100 mg l^{-1}, whereas the absorption of dissolved organic matter ranged between 1 and 4 m^{-1} at 440 nm. Measurements in 2008 were used to derive a consistent set of specific optical properties (SIOPs, see Section 3.3) through optical closure calculations using the measured subsurface irradiance reflectance and the *in situ* measurements, whereas campaigns during 2009 and 2012 served for verification and satellite data validation.

Figure 4 shows a comparison of the *in situ* field measurements and the subsurface reflectance-derived estimates, and reveals the consistency of the Secchi depth and spectrometer-derived suspended matter concentrations with the gravimetrically measured suspended matter.

3.5. Initial feasibility study

The derived set of SIOPs was tested with optical closure calculations, using image data from a range of satellite sensors. We evaluated the sensor calibration and the impact of the selected set of SIOPs by analysing the match between simulated and measured radiances and subsurface irradiance reflectance. In general, the selected set of SIOPs was found to be suitable for water-quality processing. The image data used covered both the Lower and Upper Mekong, and the sensors tested were MODIS, MERIS, Landsat ETM, SPOT 5, RapidEye, THEOS, and QuickBird. The purpose of this was to gain an understanding of the performance of the Mekong-specific SIOPs, as opposed to evaluating the potential performance of each of these sensors. A selection of the Landsat, RapidEye, SPOT 5, and ASTER data was acquired coincident with the gravimetric sampling, and a comparison of these is shown in Figure 5. Overall, these results demonstrate that the *in situ* measurements and the satellite-derived values are comparable and that the accuracy of the satellite-derived values is within an acceptable range of approximately 20% for simultaneous measurements. Note that, as is evident in Figure 5, time intervals of less than 10

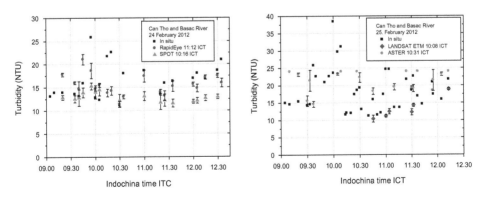

Figure 5. Comparison of *in situ* TSM as measured using gravimetric analysis (black dots on *y*-axis) and near-coincident satellite-derived turbidity (coloured symbols at *y*-axis, in NTU) using the Landsat ETM, RapidEye, SPOT 5, and ASTER sensors. Please note the satellite record time provided in the graph. The *x*-axis represents the *in situ* time of the samples taken. Measurements for both days were taken at a transect from west to east, beginning in Can Tho river at 9:00 ITC, reaching the western Bassac river bank at approximately 10:00 ITC and reaching the eastern shore at approximately 12:30 ITC.

min may cause significant differences for sites with high spatial-temporal variability such as encountered here in the Mekong. The SPOT 5 and Quickbird data required the calibration of the infrared channel to be considerably decreased in order to retrieve suitable results. This was observed for several SPOT 5 scenes. From this it was concluded that either these sensors are currently not stable for independent and operational retrieval of water constituents with a physics-based approach or more scenes were required to evaluate a systematic stable difference. For MODIS, MERIS, RapidEye, and Landsat, these initial validation and comparison findings are well aligned with multi-year valida- tion exercises within the European FRESHMON project (www.freshmon.eu) for lakes. Notwithstanding, slight adjustments of the radiometric calibration for Landsat 5, 7 and RapidEye may be introduced at a later stage, after further systematic analysis of the long- term results for these sensors.

3.6. *Operational implementation and validation*

The bulk of currently available satellite data applicable to this project are from MODIS and Landsat ETM, and these were therefore selected for further testing, processing, and operational implementation. An important consideration during this application was there- fore to determine the technical feasibility of using MODIS data to retrieve turbidity estimates for rivers with widths corresponding to 1–2 MODIS pixels.

Following these initial feasibility tests and the successful establishment of a suitable set of SIOPs as described above, five years of MODIS and Landsat data of the Mekong were uniformly processed using the Mekong-specific MIP configuration, for a total of more than 1000 MODIS scenes and 307 Landsat scenes. MODIS was used for the Mekong Delta, and Landsat was used for the Mekong Delta, Laos, Thailand, and selected areas in China (Upper Mekong), where higher spatial resolution data were required.

4. Results and discussion

Overall, the results are largely consistent in space and time. However, the technical limitations imposed by the large MODIS pixels are apparent. Not surprisingly, analysis of the MODIS scenes shows that pixels frequently contain a mixed signal of both aquatic and terrestrial targets (Figure 6).

Although the MODIS-retrieved 500 m turbidity contains quantitative information, the high variability in this measurement, due in part to the mixture of unrealistically high (terrestrial-influenced) values, precluded the use of daily per-pixel MODIS data. It was therefore decided to only extract the median value for each month, and compare this with the validation data.

In situ data suitable for validation of turbidity were collected in an autonomous station by GFZ Potsdam, over a period of two years, at a location approximately 500 m from the Bassac river at Can Tho river bridge. Additional *in situ* data in the Basac river near Can Tho were also collected by SRHMC, Ho Chi Minh City, and Hydromod, Hannover.

Figure 7 displays turbidity values as derived from MODIS, Landsat, and RapidEye as well as from the *in situ* measurements, where the MODIS values are calculated from the monthly median. It reveals that MODIS-derived turbidity follows very well the seasonal trends and, even more importantly, delivers quantitative values in good alignment with the *in situ* turbidity measurements (where highly mixed pixels were filtered out). The high- resolution satellite turbidity estimates (the value on the date where Landsat or RapidEye

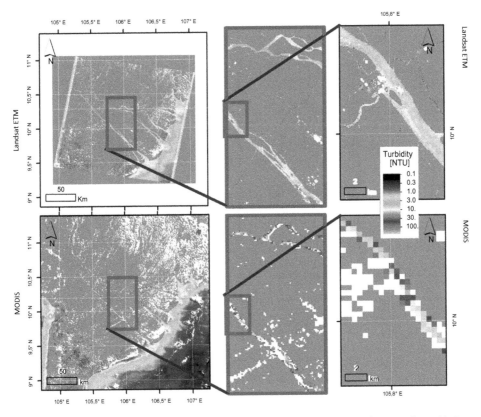

Figure 6.   Comparison of MODIS Terra 500 m and Landsat ETM 30 m imagery from 11 June 2010.

data were available) also follow the seasonal trend well, but there is an underestimation of turbidity at the lower end of the range (Figure 7).

The seasonal trends, with a southern Monsoon bringing excessive rainfall in the June–November period and a dry season lasting approximately from December to April, are captured by the intra-annual oscillations of the turbidity time series shown in Figure 5. Increasing, heavy rainfall is normally directly correlated with increased erosion, which in turn results in increased sediment loads in the rivers. These increased sediment loads are expressed as a heightened turbidity signal, which can indeed occur towards the second half of each year.

Figure 8 displays the selected turbidity maps derived from Landsat data. Broadly, the turbidity levels in the delta are the most elevated in the September scene, again, coinciding with the season for the Southern Monsoon. The onset of the dry season in December appears to be reflected in decreasing overall turbidity levels in the rivers, and the dry season March image features the overall lowest levels. Interestingly, the coastal waters reveal a different pattern, with the highest turbidity levels occurring in the January–March images. This cannot be attributed to river input, as the March image shows that this period has the lowest upstream turbidity levels. A possible mechanism for this apparent dichotomy is the tropical wind season (east and southeast wind, known as Gió Chướng), which occurs during March and primarily affects the coastal regions with increased erosion and turbidity.

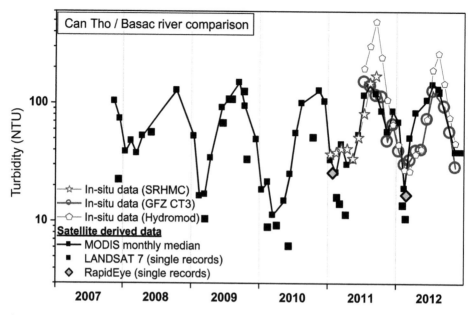

Figure 7. A comparison of monthly median MODIS, Landsat, RapidEye, and *in situ* measurements.

The above results demonstrate that the physics-based water quality remote-sensing method was successfully applied to a river system. For selected sensors, the satellite-retrieved turbidity was broadly in good agreement with both intra-day and intra-annual *in situ* measurements (Figures 3 and 5, respectively), furthermore capturing both seasonal trends and recurring patterns in river water-quality distributions (Figures 7 and 8, respectively). This suggests that the method is stable across space and time, and furthermore relatively sensor-independent (while allowing for radiometry and resolution limitations). In other words, the satellite-derived measurements can be considered as standardized, or harmonized, thereby avoiding the problems in comparing results from different research teams, equipment, measurement protocols, and across political boundaries.

In addition to the benefits of retrieving standardized measurements, the use of satellite-based data in a system such as described here also has the advantage of being very cost-effective. Although it is recommended to maintain a limited number of *in situ* measuring stations to verify satellite data calibrations, monitor any potentially significant changes in SIOPs, and calibrate satellite-derived turbidity to TSM, the deployment of a remote-sensing-based system means that significantly less of the relatively expensive *in situ* monitoring station equipment, staff, and maintenance are required.

As shown, the implemented system allows for near-real-time delivery of water-quality information as well as the analysis of time series using historical archives of satellite imagery data. This is made possible by the EWS, which can automatically process the most recent imagery data as it arrives from the receiving station, as well as batch process large data archives. These capabilities, together with the spatial and temporal stability of the output, allow for a wide range of analyses and applications, spanning from research to more applied, monitoring-type uses. Some of these potential types of analyses have been illustrated in the above results section, but the focus of this article is to present the approach itself, its implementation, and its capabilities. A more

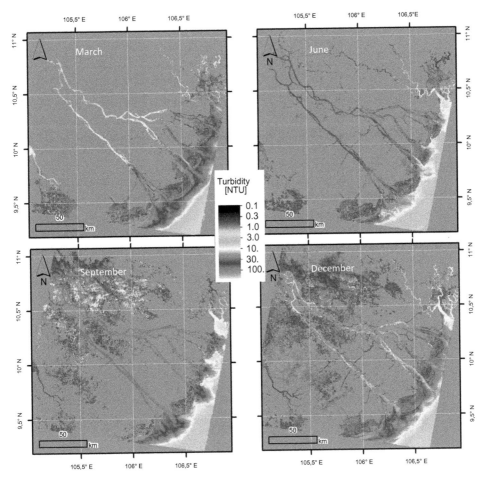

Figure 8. Selected turbidity maps derived from Landsat data temporally integrated from 2007 to 2012, for the months of March, June, September, and December.

in-depth presentation of results interpretations and potential applications is therefore beyond the scope of this article.

A key component in implementing this system was establishing SIOPs representative of the Mekong waters. These were verified as being suitable for the currently implemented algorithms through optical closure calculations using *in situ* remote-sensing measurements as well as satellite-based remote-sensing measurements, as demonstrated in Figures 2, 3, and 5. In the event that the water-quality parameter retrievals become less accurate in certain regions, perhaps indicating that the optical properties of the waters are no longer adequately represented by the current SIOPs, these SIOPs may need to be re-assessed. This would require fieldwork or a new approach to change the spectral dependency of defined SIOPs dynamically.

As previously shown, RapidEye and Landsat ETM-derived values differed from MODIS. This is to be expected for the relatively low signal-to-noise levels associated with water targets, as these two sensors are radiometrically not as stable or sensitive as MODIS. However, these remaining systematic differences might be corrected by a slight modification of the calibration values after a more detailed analysis of a larger number of scenes.

5. Conclusion

The Mekong River system, of critical importance to over 200 million people, is increasingly under pressure from a range of sources, and remote-sensing technology has the potential to contribute significantly to the need for monitoring the environmental status of the entire Mekong river system. However, such a monitoring system has to date not been implemented for inland waters at these spatial and temporal scales.

The multi-sensor, physics-based remote-sensing system presented here is completely automated, and currently operational on EOMAP processors. The Mekong-specific configuration has been implemented and is currently running, and may be unique in its application to a river system. As new satellite sensor data become available, such as from the recently launched Landsat 8 and from the upcoming Sentinel 2, 3 (scheduled for 2014), these can be assessed for incorporation into the system. Such additional sensors will further increase the temporal and spatial resolution, and support a sustainable long-term monitoring system.

Acknowledgements

We are grateful for the participation and support of Ms Trinh Thi Long, Dr Tri Vo Khac, Nguyen Minh Trung, and colleagues of the Southern Institute of Water-related Research SiWRR, and S. Gebhard, J. Huth (German Aerospace Center), in support and ground-level measurements in Vietnam, and Dr Apel of GFZ, Potsdam, Mr Giam of SRHMC, Ho Chi Minh City, and Mr Post of Hydromod for providing ground truth measurements.

Funding

The research undertaken for this article was funded by the German Ministry of Education and Research, BMBF, in the frames of WISDOM project [grant number FKZ 0330777G], [grant number 033L040].

References

Abreu, L. W., and G. P. Anderson. 1996. *The MODTRAN 2/3 Report and LOWTRAN 7 Model.* Contract, 19628(91-C), 0132.

Albert, A., and C. D. Mobley. 2003. "An Analytical Model for Subsurface Irradiance and Remote Sensing Reflectance in Deep and Shallow Case-2 Waters." *Optics Express* 11 (22): 2873–2890. doi:10.1364/OE.11.002873.

Asian Development Bank. 2012. *Greater Mekong Subregion Economic Cooperation Program,* 16pp. Manila: Asian Development Bank. http://www.adb.org/sites/default/files/gms-overview.pdf

Brezonik, P., K. D. Menken, and M. Bauer. 2005. "Landsat-Based Remote Sensing of Lake Water Quality Characteristics, Including Chlorophyll and Colored Dissolved Organic Matter (CDOM)." *Lake and Reservoir Management* 21 (4): 373–382. doi:10.1080/07438140509354442.

Buiteveld, H., J. H. M. Hakvoort, and M. Donze. 1994. "The Optical Properties of Pure Water." In *Proceedings of SPIE 2258, Ocean Optics XII,* October 26, 174. doi:10.1117/12.190060.

Bulgarelli, B., V. Kisselev, and L. Roberti. 1999. "Radiative Transfer in the Atmosphere-Ocean System: The Finite Element Method." *Applied Optics* 38: 1530–1542. doi:10.1364/AO.38.001530.

Di'az-Delgado, R., I. Ameztoy, J. Cristo'bal, and J. Bustamante. 2010. "Long Time Series of Landsat Images to Reconstruct River Surface Temperature and Turbidity Regimes of Guadalquivir Estuary." In *Proceedings of the 2010 IEEE International Geoscience & Remote Sensing Symposium (IGARSS 2010)*, July 25–30, 233–236. Honolulu, HI: IEEE Geoscience and Remote Sensing Society. ISBN: 978-1-4244-9564-1.

Doerffer, R., and H. Schiller. 2007. "The MERIS Case 2 Water Algorithm." *International Journal of Remote Sensing* 28 (3–4): 517–535.

Duan, H., R. Ma, S. G. H. Simis, and Y. Zhang. 2012. "Validation of MERIS Case-2 Water Products in Lake Taihu, China." *GIScience & Remote Sensing* 49 (6): 873–894. doi:10.2747/1548-1603.49.6.873.

Evans, R. D., K. L. Murray, S. N. Field, J. A. Moore, G. Shedrawi, B. G. Huntley, P. Fearns, M. Broomhall, L. I. McKinna, and D. Marrable. 2012. "Digitise This! A Quick and Easy Remote Sensing Method to Monitor the Daily Extent of Dredge Plumes." *PLoS One*, 7.12, http://www.ncbi.nlm.nih.gov/pmc/articles/PMC3519868/

Güttler, F. N., S. Niculescu, and F. Gohin. 2013. "Turbidity Retrieval and Monitoring of Danube Delta Waters Using Multi-Sensor Optical Remote Sensing Data: An Integrated View from the Delta Plain Lakes to the Western–Northwestern Black Sea Coastal Zone." *Remote Sensing of Environment* 132: 86–101. doi:10.1016/j.rse.2013.01.009.

Heege, T. 2000. "Flugzeuggestützte Fernerkundung von Wasserinhaltsstoffen am Bodensee." PhD thesis, Freie Universität Berlin, published as Research Report Vol. 2000-40 by Deutsches Zentrum für Luft- und Raumfahrt, Köln.

Heege, T., and J. Fischer. 2004. "Mapping of Water Constituents in Lake Constance Using Multispectral Airborne Scanner Data and a Physically Based Processing Scheme." *Canadian Journal of Remote Sensing* 30 (1): 77–86. doi:10.5589/m03-056.

Heege, T., C. Häse, A. Bogner, and N. Pinnel. 2003. "Airborne Multi-spectral Sensing in Shallow and Deep Waters." *Backscatter*, 17–19, 1/2003.

Heege, T., V. Kiselev, D. Odermatt, J. Heblinski, K. Schmieder, T. V. Khac, and T. T. Long. 2009. "Retrieval of Water Constituents from Multiple Earth Observation Sensors in Lakes, Rivers and Coastal Zones." In *Geoscience and Remote Sensing Symposium, 2009 IEEE International, IGARSS 2009*, Vol. 2, July 12–17, II-833–II-836. doi:10.1109/IGARSS.2009.5418222.

Heege, T., D. Odermatt, and V. Kiselev. 2007. "How Can I Map Water Constituents Concentrations." In *Earth Observation for Wetland Inventory, Assessment and Monitoring*, edited by N. C. Davidson and C. M. Finlayson, Vol. 17.3, 219–228. Athens: Aquatic Conservation: Marine and Freshwater Ecosystems.

Hellweger, F. L., W. Miller, and K. S. Oshodi. 2007. "Mapping Turbidity in the Charles River, Boston Using a High-Resolution Satellite." *Environmental Monitoring and Assessment* 132 (1–3): 311–320.

Irvine, K. N., J. E. Richey, G. W. Holtgrieve, J. Sarkkula, and M. Sampson. 2011. "Spatial and Temporal Variability of Turbidity, Dissolved Oxygen, Conductivity, Temperature, and Fluorescence in the Lower Mekong River–Tonle Sap System Identified Using Continuous Monitoring." *International Journal of River Basin Management* 9 (2): 151–168. doi:10.1080/15715124.2011.621430.

Kiselev, V., L. Roberti, and G. Perona. 1995. "Finite-Element Algorithm for Radiative Transfer in Vertically Inhomogeneous Media: Numerical Scheme and Applications." *Applied Optics* 34: 8460–8471. doi:10.1364/AO.34.008460.

Kobryn, H., K. Wouters, L. Beckley, and T. Heege. 2013. "Ningaloo Reef: Shallow Marine Habitats Mapped Using a Hyperspectral Sensor." *PLoS One*, X.Y (accepted June 2013).

Kuenzer, C., I. Campbell, M. Roch, P. Leinenkugel, V. Q. Tuan, and S. Dech. 2012. "Understanding the Impact of Hydropower Developments in the Context of Upstream–Downstream Relations in the Mekong River Basin." *Sustainability Science* 8 (4): 565–584.

Le, C. F., Y. M. Li, Y. Zha, D. Sun, and B. Yin. 2009. "Validation of a Quasi-Analytical Algorithm for Highly Turbid Eutrophic Water of Meiliang Bay in Taihu Lake, China." *IEEE Transactions on Geoscience and Remote Sensing* 47 (8): 2492–2500. doi:10.1109/TGRS.2009.2015658.

Malthus, T. J., E. L. Hestir, A. G. Dekker, and V. E. Brando. 2012. "The Case for a Global Inland Water Quality Product." In *IEEE International Geoscience and Remote Sensing Symposium (IGARSS)*, July 22–27, 5234–5237. IEEE. doi:10.1109/IGARSS.2012.6352429.

Martinez, J. M., J. L. Guyot, N. Filizola, and F. Sondag. 2009. "Increase in Suspended Sediment Discharge of the Amazon River Assessed by Monitoring Network and Satellite Data." *Catena* 79 (3): 257–264. doi:10.1016/j.catena.2009.05.011.

Mekong River Commission (MRC). 2010. *State of the Basin Report.* http://www.mrcmekong.org/assets/Publications/basin-reports/MRC-SOB-report-2010full-report.pdf

Odermatt, D., C. Giardino, and T. Heege. 2010. "Chlorophyll Retrieval with MERIS Case-2-Regional in Perialpine Lakes." *Remote Sensing of Environment* 114 (3): 607–617. doi:10.1016/j.rse.2009.10.016.

Ohlendorf, S., A. Müller, T. Heege, S. Cerdeira-Estrada, and H. T. Kobryn. 2011. "Bathymetry Mapping and Sea Floor Classification Using Multispectral Satellite Data and Standardized Physics-Based Data Processing." *Proceedings of SPIE, Remote Sensing of the Ocean, Sea Ice, Coastal Waters, and Large Water Regions 2011* 8175: 817503-1. doi:10.1117/12.898652.

Olmanson, L. G., P. L. Brezonik, and M. E. Bauer. 2011. "Evaluation of Medium to Low Resolution Satellite Imagery for Regional Lake Water Quality Assessments." *Water Resources Research* 47 (9). doi:10.1029/2011WR011005.

Schroeder, T., I. Behnert, M. Schaale, J. Fischer, and R. Doerffer. 2007. "Atmospheric Correction Algorithm for MERIS above Case-2 Waters." *International Journal of Remote Sensing* 28 (7): 1469–1486. doi:10.1080/01431160600962574.

Sipelgas, L., U. Raudsepp, and T. Kõuts. 2006. "Operational Monitoring of Suspended Matter Distribution Using MODIS Images and Numerical Modelling." *Advances in Space Research* 38 (10): 2182–2188. doi:10.1016/j.asr.2006.03.011.

Smith, R. C., and K. S. Baker. 1981. "Optical Properties of the Clearest Natural Waters (200–800 nm)." *Applied Optics* 20 (2): 177–184. doi:10.1364/AO.20.000177.

SWBD. 2005. *Shuttle Radar Topography Mission Water Body Data Set.* Accessed 2010. www2.jpl.nasa.gov/srtm/

Index

Milton Keynes UK
Ingram Content Group UK Ltd.
UKHW051854071024
449327UK00025B/1954